Addressing Emerging Trends to Support the Future of Criminal Justice

Findings of the Criminal Justice Technology Forecasting Group

John S. Hollywood, Dulani Woods, Andrew Lauland, Brian A. Jackson, Richard Silberglitt

For more information on this publication, visit www.rand.org/t/RR1987

Library of Congress Cataloging-in-Publication Data is available for this publication.
ISBN: 978-0-8330-9905-1

Published by the RAND Corporation, Santa Monica, Calif.

Cover: Getty Images/imaginima

Preface

The Bureau of Justice Assistance has established the Criminal Justice Technology Forecasting Group to assess the effects that major technology and social trends could have on criminal justice, including both threats and opportunities, and identify potential responses. This report reflects recent discussions and findings of the group, which include identifying several dozen trends, their effects, and more than a dozen proposed responses. The bureau sponsored the work reported here. The report should be of interest to criminal-justice practitioners, developers, funders, and policymakers seeking an assessment of emerging trends and how the criminal-justice community might best take advantage of the resulting opportunities and mitigate the resulting threats.

RAND Justice Policy

The research reported here was conducted in the RAND Justice Policy Program, which spans both criminal and civil justice system issues with such topics as public safety, effective policing, police–community relations, drug policy and enforcement, corrections policy, use of technology in law enforcement, tort reform, catastrophe and mass-injury compensation, court resourcing, and insurance regulation. Program research is supported by government agencies, foundations, and the private sector.

This program is part of RAND Justice, Infrastructure, and Environment, a division of the RAND Corporation dedicated to improving policy- and decisionmaking in a wide range of policy domains, including civil and criminal justice, infrastructure protection and homeland security, transportation and energy policy, and environmental and natural resource policy.

Questions or comments about this report should be sent to the project leader, John S. Hollywood (John_Hollywood@rand.org). For more information about RAND Justice Policy, see www.rand.org/jie/justice-policy or contact the director at justice@rand.org.

Contents

Figures and Tables

Figures

Tables

Summary

The Bureau of Justice Assistance (BJA) has established the Criminal Justice Technology Forecasting Group (CJTFG), an expert advisory panel that includes both practitioners and researchers to deliberate on the effects that major technology and social trends could have on criminal justice in the next three to five years and identify potential responses. This report captures the results from the CJTFG's meetings and initiatives. It presents the emerging trends and highlights of the group's discussion about them. It then presents the results of analyses to assess connections between the trends, leading to recognizing the crosscutting themes that those connections represent. In addition, the report presents analyses to generate a set of recommended ways to address the trends that the full group reviewed and approved.

The CJTFG covered a wide range of topics in its deliberations, identifying close to two dozen trends contributing to six overarching themes along with their potential impacts. The group, with the assistance of the RAND Corporation, BJA, and the Institute for Intergovernmental Research, have identified more than a dozen ways ahead in response and sponsored initiatives in direct support of four of these ways ahead. Table S.1 summarizes the themes, constituent trends, and ways ahead.

Business processes for new technologies are lacking. CJTFG members described a rich vein of governance, business-process reengineering, standardization, and silo-breaking that must be addressed to leverage new technologies successfully. Business cases are part of larger operationalization processes; the group discussed how cost assessments, requirement management, governance mechanisms, and security, privacy, and civil-rights protections need to be included as well. The CJTFG called for both business cases and processes to be developed; to start, the CJTFG and the Global Justice Information Sharing Initiative have cosponsored development of core information exchanges between agencies providing services to offenders and the rest of the criminal-justice system, as well as exchanges between medical and criminal justice agencies.

There are opportunities in big data and analytics but challenges to leverage them. Opportunities include using data and analytics to identify locations, times, and people at increased risk of crime; risk-based bail and sentencing support; and tools to identify services needed by those on community supervision. The emergence of situational-awareness displays in improving form factors (e.g., tablets, smartphones) offers opportunities to get needed information to practitioners. The CJTFG called for BJA to sponsor visits to learn from agencies pioneering the new technologies.

However, the CJTFG noted the lack of awareness of references on criminal-justice technologies as a barrier for agencies to procure and use them effectively. Resources are difficult to locate and understand, especially for newcomers. Technical and organizational barriers to share

Table S.1
Themes, Trends, and Ways Ahead from the Criminal Justice Technology Forecasting Group

Theme	Trend	Way Ahead	CJTFG Action
Lack of business cases and processes for new technologies[a]	• Lack of core business cases and other key reference material for new technologies • Need to define and disseminate the value of fusion centers • Lack of established business processes for operationalizing new technologies	• Develop business cases for key technologies. • Develop common business processes for operationalizing key technologies.	The CJTFG cosponsored specifying data exchanges for those that provide services to offenders as well as exchanges between medical and justice agencies (see Appendix E).
Emergence of big data and analytics and challenges using them[a]	• Lack of awareness of existing training and reference material on criminal-justice technologies • Emergence of analytics and enabling big data • Emergence of situational-awareness displays, enabling devices, and data streams • Increasing pressures to investigate cybercrimes • Need to advance from small-scale information-sharing to nationwide information-sharing	• Conduct research and experimentation on improving how technology resource materials are made available to both practitioners and researchers. • Through site visits and interviews, learn from agencies how they are using emerging data technologies. • Develop a federally sponsored crime-analysis capability.	The CJTFG recommended that BJA directly sponsor a learning tour on agencies' use of new data technologies. The CJTFG authored a white paper on developing a crime analysis capability (see Appendix C).
Security, privacy, and civil-rights challenges[a]	• Increasing pressures to have cybersecurity protections • Increasing pressures to address issues that consistently affect agencies' use of new surveillance technologies • Lack of legal foundations and case law for new surveillance technologies • "Going dark"—unbreakable commercial encryption is hampering criminal investigations	• Incorporate security, privacy, and civil-rights protections throughout the previously recommended common business processes. • Develop materials to educate the public on how criminal-justice technologies work (or do not work) in the real world. • Collect hard data on the extent of the "going-dark problem" and investigate work-arounds.	
Getting to true, field-wide information integration	• Need for integration to enable new models of criminal justice across the enterprise • Reality of "have and have-not" agencies being a barrier to shared criminal-justice capabilities • Need to support digital-evidence management on a massive scale	• Research methods to change cultures to support information-sharing and safeguarding. • Require the exportability of core criminal-justice record–system data. • Develop regional models for information-sharing capabilities.	The CJTFG cosponsored a resolution on this topic (Appendix F, available online).

Table S.1—Continued

Theme	Trend	Way Ahead	CJTFG Action
Improving safety and community relations	• Increasing pressures to move toward guardianship, with competing pressures to crack down on violence and terrorism • Increasing pressures for law enforcement to focus on accountability • Need to support fielding of body-worn cameras on a large scale • Need for less-lethal weapons to reduce the number of lethal use-of-force incidents	• Identify combinations of practices and technologies that offer the greatest potential in reducing crime and improving community relations. • Facilitate exchanges between the United States and the United Kingdom on lessons learned on using cameras for both investigative and accountability purposes. • Develop new immobilization and restraint technologies.	
New technologies and new consequences	• Fact that technology can have both unintended consequences and unanticipated benefits • Emergence of touch- and rapid-DNA systems • Emergence of remote weapon-detection capabilities	• Have the core business cases and processes include risk-assessment elements to mitigate unanticipated consequences. • Assess the potential of remote weapon-detection capabilities.	

NOTE: DNA = deoxyribonucleic acid. *Less-lethal* has a specific meaning in law enforcement and refers to the fact that such use of weapons as conducted-energy weapons can result in death.

[a] This theme is part of an overall narrative: Business-process obstacles and challenges in ensuring security and civil liberties are hampering information technology (IT) opportunities.

information also hamper efforts to use data, especially across organizational lines. (Below, we discuss information-sharing barriers.)

There are challenges to ensure security, privacy, and civil-rights protections. Without planning and engagement on these issues, agencies can use systems in ways that create community-relations problems and can even put other criminal-justice agencies' access to needed technologies at risk. The CJTFG called for building security, privacy, and civil-rights planning and protections into the business cases and processes recommended in Table S.1.

There is a need to advance from small information-sharing pilots to field-wide integration of information. Group members discussed a variety of pilots, standard-development efforts, and other initiatives to share information, but progress toward nationwide integration has consisted largely of small-scale interagency sharing and a few nationwide initiatives.

The group noted that organizational culture is commonly presented as a major barrier to improved information-sharing and safeguarding. Rather than treat it as an insurmountable barrier, the CJTFG called for research on changing it to better support information-sharing.

One condition for nationwide sharing is for agencies to be able to use and transfer their own record data freely. This in part requires record-management systems to support exporting data in easy-to-translate formats. CJTFG members noted that, if agencies had these capabilities, searches and information-sharing could be an order of magnitude more extensive than they are now, and they cosponsored a resolution to mandate these capabilities with Global.[1]

There are increasing and competing pressures to improve public safety and community relations. Agencies face pressures to minimize use of force, yet crack down on violence and terrorism at the same time. Penalties for making a mistake in responding to these pressures can be grave. Group members noted that a video of an apparently unjustified police shooting can affect perceptions of every police officer in the country, while a mass shooting can drive widespread fears of classes of people. Both have occurred together; police who were caught on video shooting Laquan McDonald in Chicago, Illinois, and the mass shooting at the Inland Regional Center in San Bernardino, California, were major stories in December 2015. To help with this problem, the group called for research to identify combinations of practices and technologies that offer the greatest potential in reducing crime and improving community relations together.

Technologically, the group recommended research and development to improve less-lethal technologies to subdue and restrain subjects. Group members discussed prior lethal-force incidents in which, if a suitable less-lethal option had been available, it likely would have been used, at much less cost to police–community relations and to life.

Overall, business-process obstacles and challenges in ensuring security and civil liberties hamper IT opportunities. When we look across all the themes, we find a narrative linking half of them: There are promising opportunities to use IT to improve law enforcement, but taking advantage of them is hampered by a lack of clear and compelling business cases, a lack of established processes for implementing technologies, a lack of nationwide sharing capabilities to get needed information, and problems with ensuring security, privacy, and civil-rights protections. These issues represent challenges for a single criminal-justice agency; integration across agencies is especially challenged. Making progress in the proposed ways ahead for these themes would help criminal-justice agencies leverage IT opportunities more effectively.

---

1 This would move the U.S. Department of Justice toward requiring that any agency's use of department grant money to buy record-management systems would require that those systems comply with this data-exportability requirement.

Acknowledgments

We would like to thank the members of the Bureau of Justice Assistance's Criminal Justice Technology Forecasting Group (CJTFG) for their participation and assistance with this effort. We would also like to acknowledge the support and contributions of Kristen Mahoney, James P. McCreary, John Markovic, and Michael Roosa of the Bureau of Justice Assistance. We would also like to acknowledge the valuable contributions made by the peer reviewers for this report, Julie Wartell of the Analysis Group and Isaac R. Porche III of RAND.

We would also like to thank Terri Pate, Gina Hartsfield, Christina M. Abernathy, Mary Jo Dodd, and Donna Lindquist of the Institute for Intergovernmental Research for their assistance in providing administrative and logistical support to the CJTFG, as well as the Critical Nontraditional Information Exchange (CNIE) Task Team, jointly sponsored by the CJTFG and the Global Justice Information Sharing Initiative. We would also like to thank the members of the CNIE team for their work on defining information exchanges needed to improve outcomes for offenders and enhance public safety as a result.

Abbreviations

ARJIS	Automated Regional Justice Information System
BJA	Bureau of Justice Assistance
BWC	body-worn camera
CAD	computer-aided dispatch
CISA	Cybersecurity Information Sharing Act
CJIS	Criminal Justice Information Services
CJTFG	Criminal Justice Technology Forecasting Group
DHS	U.S. Department of Homeland Security
DNA	deoxyribonucleic acid
DOJ	U.S. Department of Justice
FBI	Federal Bureau of Investigation
IACP	International Association of Chiefs of Police
IoT	Internet of things
IT	information technology
LAPD	Los Angeles Police Department
LInX	Law Enforcement Information Exchange
LPR	license-plate reader
N-DEx	National Data Exchange
NIBRS	National Incident Based Reporting System
NIEM	National Information Exchange Model
UCR	Uniform Crime Reporting

Discussion and Findings

Overview

The Bureau of Justice Assistance (BJA) established the Criminal Justice Technology Forecasting Group (CJTFG) to identify and assess the major technology trends, along with corresponding opportunities and threats, which are most likely to affect criminal-justice practice significantly in the next three to five years. The CJTFG was tasked to (1) identify major, emerging social and technology trends; (2) assess the impacts that these trends could have criminal justice, in terms of both threats and opportunities; and (3) identify promising responses to the trends. The latter include recommendations that

- advise agencies on what they can do to use key technologies more effectively
- advise technology providers and funders on what they can do to make technologies more useful and less risky to implement
- are intended to create deliverables that directly support the criminal-justice field.

This report captures the results from the CJTFG's meetings and initiatives.[1] It presents the emerging trends and highlights of the group's discussion about them. It then presents the results of analyses to assess connections between the trends, leading to recognizing the cross-cutting themes that those connections represent. In addition, the report presents analyses to generate a set of recommended ways to address the trends that the full group reviewed and approved. Several of the recommended ways ahead resulted in initiatives that the CJTFG cosponsored; we capture the deliverables from these efforts in the appendixes.

Methodology

The CJTFG met four times to address the tasks as defined above. The first three meetings were in person; the final meeting was held virtually. In Appendix B, we provide the process, methodologies, analyses, and analysis results.

[1] In terms of timing, the second CJTFG meeting occurred after the Ferguson, Missouri, shooting of Michael Brown and subsequent protests but before controversies of over law-enforcement operations peaked with the Baltimore, Maryland, death in custody of Freddie Gray and subsequent protests and riots. President's Task Force on 21st Century Policing, 2015, also had not been released. The result is that the group discussed the issues that these events raised at the first two meetings but not to the extent that it has since the spring and summer of 2015. The third meeting addressed these issues (mostly in the context of emerging external pressures shaping criminal justice) in some detail.

Meetings 1 (2014) and 2 (2015) featured a combination of guest speakers on technology and criminal-justice topics, followed by focused breakout groups. Meeting 1, held in March 2014, included panels and presentations on current technology initiatives, including those highlighting initiatives by the National Institute of Justice, the Global Advisory Committee, BJA, the International Association of Chiefs of Police (IACP) Technology Center, Regional Information Sharing Systems, and the program manager for the Information Sharing Environment's initiatives on enabling the future of information-sharing. Panelists also discussed and rank-ordered technologies and technology issues by how much impact they would likely have on criminal justice in the next three to five years (i.e., 2015 to 2020).

Meeting 2's breakout topics included the operationalization of mobile technologies; sharing court and corrections data with law enforcement; real-time crime centers and fusion centers; and criminal identity–management improvements.[2] The meeting also saw special presentations on from the Los Angeles Police Department (LAPD) (which also hosted the meeting) on the future of the deployment of body-worn cameras (BWCs) and from the program manager for the Information Sharing Environment on preventing and responding to cyberattacks.

Meeting 3 (2016) was a structured discussion to further develop the trends and their likely impacts on criminal justice. The group also started to develop potential responses to the problems and opportunities that the trends raised. To develop the list of trends, we asked the group to consider technology and related social and political trends. We asked members to focus on trends well underway, as opposed to technologies just emerging now, in line with BJA's three- to five-year horizon. We also asked members to consider the effects of interactions between trends. To develop the impacts, we asked panelists to assess the trends' likely effects on criminal-justice operations in the next two to five years, including both opportunities and threats.

To develop potential responses, we asked members to consider ways to advise agencies on addressing trends more effectively and ways to advise developers and funders on actions that would make technologies more useful—or less of a threat. We asked members to identify any known evidence on how potential solutions have worked to date (i.e., results of field experiments) or at least logical requirements for future development of such evidence (e.g., benchmarks or frameworks for defining, measuring, and tracking success).

The group discussed trends in several areas, including information-sharing and safeguarding, video and other sensor data, digital forensics, operational technologies, and "anything not previously covered." Each topical discussion began with brief "fire-starter" discussions to provide examples of new technological and educational developments; these included presentations on the IACP's Law Enforcement Cyber Center (Law Enforcement Cyber Center, undated [a]); the Open Source Policing resource portal (Open Source Policing, undated); predictive policing about people identified as being of high risk of being a party to violence (based on efforts undertaken in Chicago); BJA's Body-Worn Camera Toolkit resource portal (BJA, undated); findings from the "Data and Civil Rights: A New Era of Policing and Justice" conference in 2015 (Data and Civil Rights, 2015); and the graduated-reentry concept for reducing incarceration rates.

Our approach to capturing the deliberations was based on the Futures Wheel concept (Glenn, 1972), which captures an initial trend, potential direct (first-order) impacts from

---

[2] The group covered the last item in the session "Removing Offender Anonymity to Better Deploy Resources by Overlaying Data from Warrants, Probation/Parole, Juvenile Justice, and Social Services."

that trend, potential subsequent indirect (second-order) impacts following from the first-order effects, and the links between them, in an expanding radial diagram (Figure 1.1). We expanded the Futures Wheel concept to capture trends, trend details, impacts (opportunities and threats), potential responses, and key relationships between them.

We used a mind-mapping tool to capture the specific points of the CJTFG's deliberations in near-real time during meeting 3. A mind map is a diagram that captures information in a hierarchical format, like an expanding tree, showing how the information can be organized into levels of groups (e.g., branches, leaves), as well as how different pieces of information relate to each other (see, for example, Pinola, 2013). Using the mind-mapping tool allowed the facilitator to capture the discussion dynamically as it introduced and moved around different trends, implications, and potential responses, including writing details of structured relationships for each trend (i.e., links between trends, implications, potential responses, and important details about each) and relationships between trends. Group members saw the mind map as it was being created and were able to make additions and edits directly. Following the meeting, the mind map allowed us to view large portions of the discussion results at once and to move discussions of themes and recognize and draw additional relationships. The radial, continuously expanding structure of a mind map was also a natural fit for the Futures Wheel approach.

Figure 1.2 shows visualizations of the original Futures Wheel (as developed by Glenn) as we implemented it with the mind-mapping tool.

Figure 1.1
A Simple Futures Wheel

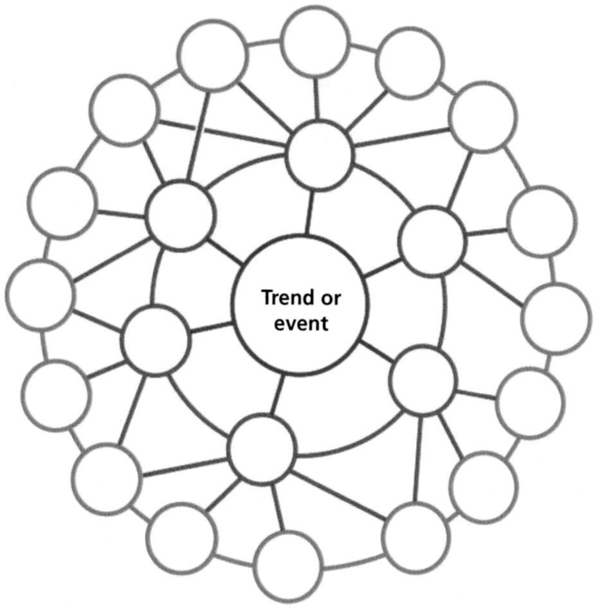

SOURCE: Zapyon, 2009. Creative Commons Attribution 2.5 Generic license.
NOTE: The blue center is the first-order effect. Red indicates a second-order effect, and so on.
RAND *RR1987-1.1*

Figure 1.2
Implementing an Expanded Futures Wheel as a Mind Map

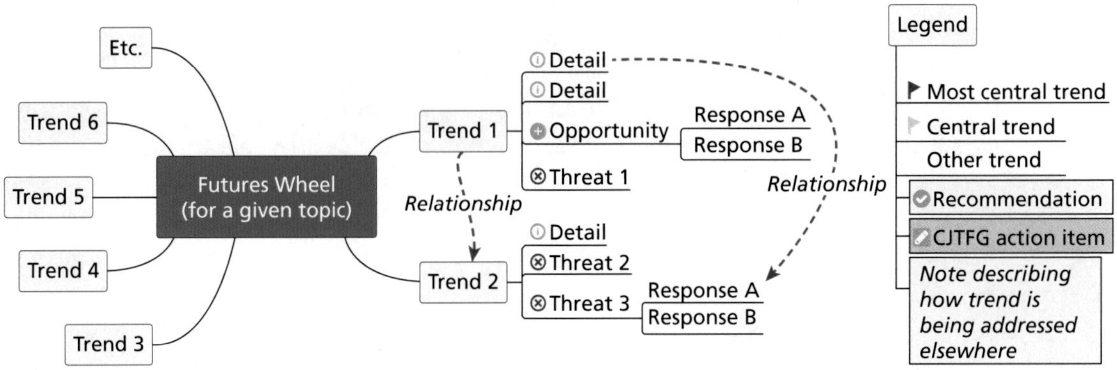

RAND *RR1987-1.2*

At the end of meeting 3, we had captured 22 trends and associated impacts, potential responses, and relationships among them. Appendix D, available online, provides detailed notes from the meeting; the final mind map is available as a poster, also available online.

Analyzing the Trends, Impacts, and Responses, Post–Meeting 3

Following the meeting, we reviewed the complete mind map of trends generated during the meeting, along with notes on meetings 1 through 3, to identify relationships between the trends.

Once we had a set of trends and relationships between them, we assembled them into a "social network" using the open-source social-network analysis tool, NodeXL (NodeXL, undated). We performed three analyses on the trends and relationships between them. First, we visualized the full network of interdependencies (i.e., how the trends relate to one another).

Second, we consolidated the 22 trends into six overarching *themes* on how technologies are affecting criminal justice, based on how densely connected the themes were (i.e., themes with lots of relationships between them were combined into the same theme). This helps elevate understanding by going from *what kinds of trends existed* to *what the trends were collectively trying to tell us.*

Third, we identified the trends having the most connections to other trends, either directly or indirectly, making them the *most centrally involved in driving impacts on criminal justice.* (See Appendix B for the discussion of what "central" means, mathematically.)

Post–Meeting 3 Action Items

In the weeks following CJTFG meeting 3, BJA, in conjunction with the RAND team, identified several near-term action items for the CJTFG to respond to some of the pressing trends and impacts. Two task teams were created to improve sharing of information about offenders with external service providers, cosponsored by the U.S. Department of Justice's (DOJ's) Global Justice Information Sharing Initiative. A third task was to develop a concept of operations for a federally sponsored crime-analysis capability. In "Recommendations and Actions," we describe these further (as action items 1 through 3); work on the action items began within a few months following CJTFG meeting 3.

Identifying and Developing Recommendations

We reviewed the mind-map elements and notes from the CJTFG's meetings covering proposed responses to the emerging trends. From analyzing the discussions of potential responses, we identified a set of candidate recommendations that were

- *specific*, calling for an identified party to carry out a defined set of activities to address a problem or leverage an opportunity
- *substantive*, offering the possibility to gain significant benefits to criminal-justice practice in the next three to five years
- *comprehensive*, collectively addressing all the themes and most of the problems and opportunities recognized in the individual trends
- *nonredundant*, in terms of not being addressed through other major initiatives.

Meeting 4 (2016), a conference call, was a structured review of the analyses described above and the draft recommendations. The meeting led to consolidating and dropping some of the recommendations, refining others, and creating and endorsing two additional recommendations. The group also endorsed a resolution made to the Global Advisory Council to require data in core criminal-justice information systems to be made exportable (described below as action item 4).

Summarizing the Trends and the Overall Themes

Figure 1.3 presents a mind map of the 22 trends and six overall themes emerging from the CJTFG's deliberations and our subsequent analysis. The four most central trends are noted with red flags; the next four most central trends are marked with yellow flags.

In the rest of this section, we summarize the overall themes and the trends that make them up. Appendix D, available online, presents detailed notes on the CJTFG's discussions on each theme. We begin with the red lines in Figure 1.3, capturing a key storyline emerging from CJTFG discussions that ties the themes together. We then cover themes and trends from the top right of Figure 1.3 (starting with trends in the *lack of business cases and processes* theme) and working clockwise.

Opportunities in Information Technology Are Hampered by a Lack of Business Cases, Implementation Processes, and Security, Privacy, and Civil-Rights Knowledge

The red lines in Figure 1.3 show a key emerging story—information technology (IT) opportunities abound, but taking advantage of them is strongly challenged by lack of clear business processes and by concerns related to security, privacy, and civil rights. IT opportunities range from data collection (including "surveillance" systems, such as cameras and license-plate readers [LPRs]) to data management (e.g., cloud installations) to analytics for processing all of those data to new tools and devices (e.g., smartphones, tablets, virtual-reality goggles) for making that information available to personnel at all levels. However, there is a broad lack of knowledge on how to use and acquire the new technologies efficiently and effectively. Group

Figure 1.3
Emerging Themes and Constituent Trends from the Criminal Justice Technology Forecasting Group

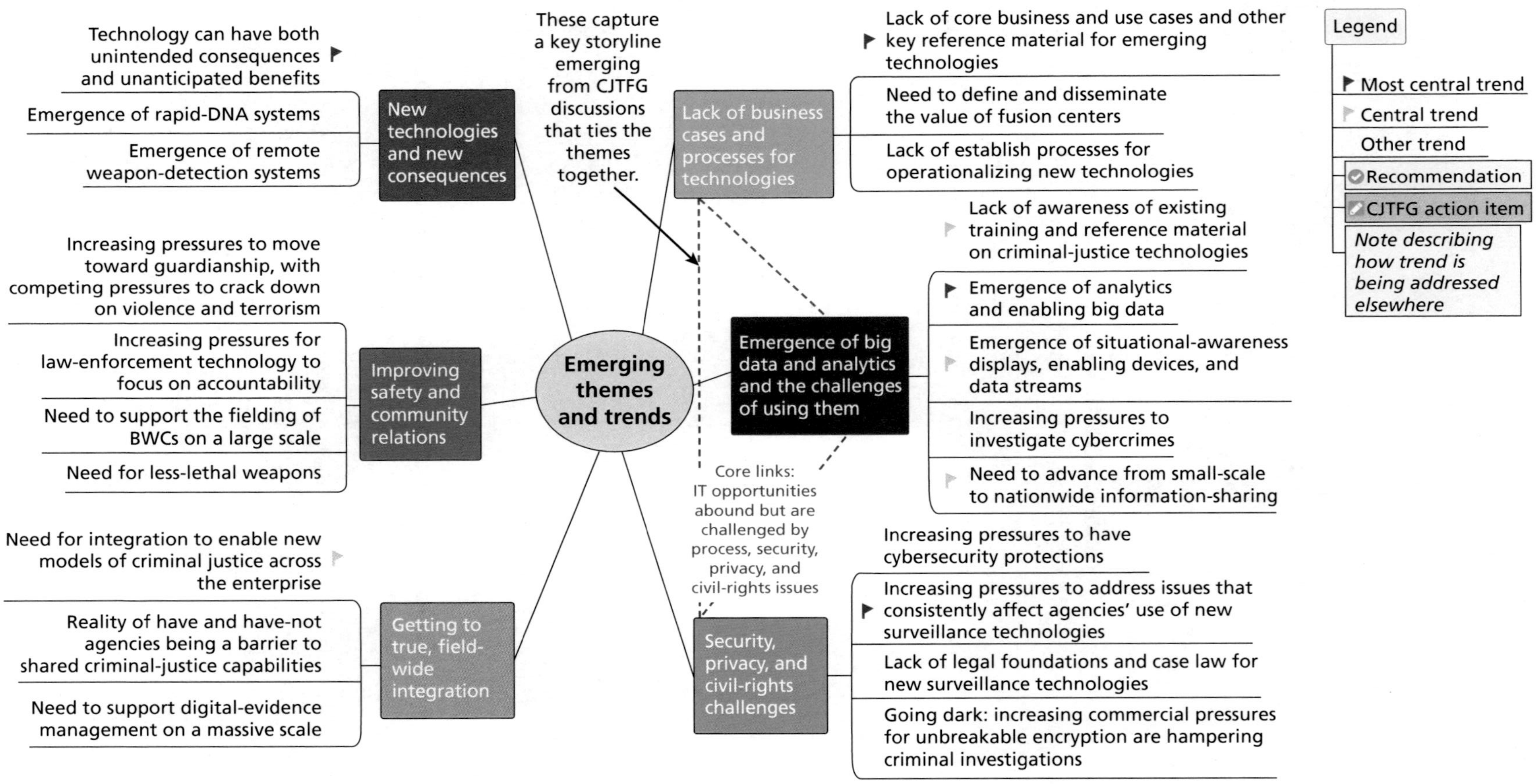

NOTE: Colored rectangles are themes. Outlined items are trends. A red flag indicates a most central trend. A yellow flag indicates a second-tier (central but not most central) trend. DNA = deoxyribonucleic acid.

RAND *RR1987-1.3*

members cited lack of knowledge of the security, privacy, and civil-rights protections needed to employ these new information technologies safely as key challenges as well.[3]

This narrative extends to other themes. Getting field-wide system integration is, in part, an IT challenge that faces significant process, security, privacy, and civil-rights challenges. Technologies for safety and community relations include BWCs and other technologies for accountability—with strong surveillance aspects—which raise a range of security, privacy, and civil-rights issues. Emerging technologies commonly include strong IT components (i.e., the information generated by touch DNA and weapon-detection sensors), but, as a result, new and unintended consequences are especially prevalent along security, privacy, and civil-rights lines, especially for surveillance-related technologies.

Business Cases and Business Processes for Technologies Are Lacking

This theme brings together the lack of core business cases for new technologies with the lack of core business processes. It negatively affects virtually all efforts to implement new technologies in support of criminal justice.

There is, as panelists described, a consistent lack of core business cases and use cases needed to inform criminal-justice technology investments. Panelists noted that a perpetual problem is letting technology and unrealistic expectations about it drive technology adoption, rather than considering what practitioners really need, and that there is a need to develop business-value propositions and use cases that better describe real operational needs for new technologies. This was discussed primarily in the context of information-sharing but applied to other technologies as well. The cases should include what benefits agencies can expect to get in terms of law-enforcement outcomes, with references to supporting evidence. The cases also need to be specific, providing checklists for agencies on what to get and how to use it in order to get the expected benefits. There is also a strong need to involve stakeholders both inside and outside of agencies in developing cases in order to get widespread buy-in.

These cases are a key requirement to address increasing civil-rights, privacy, and other policy challenges to technologies, especially those that can be labeled "surveillance." Challenges have been most acute when technologies have been fielded without a proper understanding of how those technologies will be used.

This was one of the most central trends. It hampers the effective operational use of a variety of emerging technologies; it also causes problems for surveillance-technology policy because critical "what are you planning to use it for?" policy questions can go unanswered.

A primary example of the need to develop and disseminate business cases is **the need to define and disseminate the value of fusion centers across the range of political and community stakeholders**. Panelists noted that fusion centers have provided products and services that have operational value. One example from the discussion was centers providing rapid-response crime analysis following major incidents in ways considered to be very helpful in identifying and apprehending perpetrators quickly. However, advocates as a group have been accused of failing to demonstrate fusion centers' value consistently to funders in Congress and to local law enforcement (e.g., the U.S. Senate, Committee on Homeland Security and Government Affairs, Permanent Subcommittee on Investigations highly critical report [U.S.

---

[3] Here, *privacy protections* refers to safeguards against unwarranted access, review, and use of individuals' personal data. *Civil-rights protections* refers to safeguards against technologies being used for discriminatory purposes.

Senate, 2012]). Panelists noted that individual fusion centers need to become known, consistently, for providing a specific product or service that has real operational value.

There is a lack of established business processes for operationalizing new technologies. Successfully leveraging new technology is (at least) as much about changing business processes as it is about anything else. Although the group name includes the phrase "technology forecasting," the group described a rich vein of governance, business-process reengineering, standardization, and silo-breaking that must be addressed as well. Panelists noted common challenges with technology implementation that are emerging and need to be addressed, including cost assessments, requirement management, governance mechanisms, and security, privacy, and civil-rights protections. Core business cases are an important necessary but far from sufficient component of larger operationalization processes.

Big Data and Analytics Are Emerging and Bringing Challenges with Them

This theme connects the emergence of big data and analytics, situational-awareness displays, cyber investigations, and the information-sharing needed to use them effectively. This theme also includes the general lack of awareness of technology resources and training, which hampers information-sharing and analysis.

There is a lack of awareness of existing training and reference material on criminal-justice technologies. The panel discussed numerous examples of resources and training opportunities that exist, including federally sponsored and association-sponsored material, but that are not widely known. They noted that many resources are difficult to locate and understand, especially for newcomers to a specific technology. Examples included technical information-sharing and cybersecurity and cybercrime training and references.

This was a central trend for the CJTFG. It hampers adoption and use of a variety of emerging technologies and makes those technologies vulnerable to cyberattack.

The emergence of analytics and big data offers substantial opportunities, but also substantial barriers and risks, for the criminal-justice community. The emergence of analytics and big data received the most discussion time at the third CJTFG meeting. Potential applications the panel discussed included

- predictive policing, which is the use of machine learning to identify locations, times, and people at increased risk of crime, followed by taking actions to mitigate those risks (e.g., Perry et al., 2013)
- the use of statistical models to inform bail, sentencing, and early-release decisions, as well as support matching persons to the services they need (e.g., Casey et al., 2014).

Panelists noted that a smart integration of technology, data, analytics, and good, community-based practices at a pilot site could provide an example of what good policing could look like. The panel also discussed possibilities to use data integration and decision-support tools to better integrate criminal-justice and social-service interventions.

However, there is a great deal of variation in what is being done, with little consensus on what good policing driven by analytics and data should accomplish, much less what best practices are (an example of the larger lack of core business cases for technology). There are consistent concerns related to privacy, civil rights, and community buy-in when employing analytics and big data for criminal-justice applications. Panelists also noted a general lack of knowledge of crime analysis (an example of the general lack of technological knowledge, discussed earlier) and its value in general (both traditional and newer, computer-assisted forms) and cultural resistance within agencies toward computerized, quantitative decision support. For agencies that do have analytics capabilities, panelists noted concerns about duplicating analysis efforts and the larger lack of sharing data limiting the analyses that can be done.

This is one of the most central trends. On the opportunity side, it presents opportunities for making sense out of all the new data being collected and shared across agencies. On the challenge side, concerns about the privacy and civil-rights risks that analytics pose are driving surveillance policy challenges. In addition, group members saw leveraging analytics as hampered by a lack of awareness among practitioners on what technology capabilities were available and how to access them.

The emergence of situational-awareness displays, along with enabling devices and data streams, similarly offers substantial opportunities, but also substantial barriers and risks to their implementation, for criminal justice. Panelists noted that the field is seeing novel forms of user interfaces, including improved display eyeglasses (with Google Glass mentioned as an early example) and improved voice-activated systems (with Apple Siri and Amazon Alexa mentioned as early examples).

To support the field, panelists suggested leveraging new technologies to assist officers with filling out reports (autopopulating fields, for example) to save time and reduce error.

To support operations centers and command posts, panelists discussed tailored displays and supporting data analysis. Two potential applications were discussed—real-time monitoring of the stress levels of voice communications (both into 911 call centers and over in-field networks) and real-time monitoring and control of video feeds. Panelists noted two major risks, however: human-factor problems, such as the physical burden of carrying too many devices and the cognitive burden of being overloaded with too much information, and a lack of in-house information and analytical capabilities needed to share, process, and integrate all of the data needed to populate situational-awareness screens.

This was a central theme from the CJTFG. Much like the emergence of big data and analytics, it offers major opportunities to make sense of data about the world, but there are also potential unintended consequences, such as information overload.

There is a need to advance from small-scale information-sharing to nationwide information-sharing. Panelists noted seeing progress on information-sharing, such as increasing migration to regionalization, shared services, and cloud models for information-management systems for police, courts, and corrections; computer-aided dispatch (CAD) systems; and other key criminal-justice data systems. However, panelists expressed concerns about having made limited progress on information-sharing after decades of effort.

Panelists expressed concern about repeatedly hearing the same problems with achieving information-sharing, including government, policy, cultural, and other barriers, along with the same generic solutions preferred (e.g., "calls for better leadership"). They noted that there have been many pilots, experiments, summits, and so on, but there has been only limited progress outside of small-scale interagency sharing and a few nationwide systems and initiatives. Panelists noted that it continues to be difficult to get people in the same room to share information because of a combination of inertia and unwillingness to share for a variety of reasons, including culture.

This was a central theme. Sharing data was seen as a necessary condition to be able to use data in analytics, situational-awareness displays, and so on; it is also hampered by a lack of use cases on what information really needs to be shared, awareness of how to share it, and surveillance-related policy pressures applying to sharing surveillance-related data.

Security, Privacy, and Civil Rights See Challenges from New Technologies

This theme covers pressures to address issues that consistently affect agencies' use of new technologies, especially surveillance technologies—notably, security, privacy, and civil-rights policy issues. This trend linked to every surveillance-related trend (both sensors and analytics). Panelists also discussed that specific issues include demands for cybersecurity protections and demands to tightly define how the technologies would be used (i.e., core use cases).

There are increasing pressures to have cybersecurity protections and, closely related, **there are increasing pressures to conduct cybercrime investigations**. The CJTFG devoted substantial time to cybersecurity in its meetings. On the positive side, the CJTFG heard presentations on increased federal support for cyber, including increased Federal Bureau of Investigation (FBI) support for states' cyber investigation and the establishment of small-scale pilot efforts to help agencies improve their cybersecurity efforts. BJA has sponsored the Law Enforcement Cyber Center, which provides resources to help agencies build and enhance cyber prevention, investigation, prosecution, and response policies and protocols (Law Enforcement Cyber Center, undated [a]). The CJTFG also discussed a rise in promising cybertechnologies and practices, including shared-service and cloud security models, as well as offensive, counter-hacking strategies.

On the negative side, the CJTFG heard presentations describing little unity of effort on cyber efforts. As one example, the group learned of a cybersecurity fact sheet, for example, that recommended reporting cyberattacks to eight different organizations. Many cyber efforts to date are pilot initiatives; others were useful only if an agency had computer security personnel. Members reported a lack of knowledgeable people on cyber in general, which is especially acute for criminal-justice agencies ("can't afford them, and when we do get some they tend to leave quickly," was one comment). The group heard about a lack of data on cyberattacks and crimes, with one presentation speculating up to billions in fraudulent social-service payments being made as a result of successful cyberattacks, with the lack of data meaning that the speculation is impossible to prove or disprove.

There are increasing pressures to address issues that consistently affect agencies' use of emerging surveillance technologies. A set of related challenges to using technology is emerging, not just for BWCs but for any technology with a surveillance aspect. These currently include automated LPRs, cell site simulators, touch DNA systems, field DNA-collection systems, facial recognition, health telemetry sensors, vehicle telemetry tracking, persistent Internet-of-things (IoT) sensors, and social media monitoring and analytics used by criminal-justice agencies. The common challenges involved with surveillance-technology implementation include demands for, and disputes over, specific use cases for the technologies, information-assurance policies (note the link to increasing pressures for cyber protections, discussed later), usage policies, and community and external expert participation and oversight. These can be seen as special case subsets of the general need to improve and standardize requirements and other business processes for fielding new technologies, discussed previously.

Another, somewhat competing aspect of dealing with the emerging surveillance technologies relates to public expectations. Panelists noted that the public expects product advertising and television show levels of performance (the so-called CSI effect) resulting from new technologies; expectations are often highly unrealistic.

This is one of the most central trends, complicating every surveillance technology–related trend (both sensors and analytics). Panelists also discussed how specific issues include demands for cybersecurity protections and demands to tightly define how the technologies would be used (i.e., core use cases).

There is a lack of legal foundations and case law for emerging surveillance technologies. U.S. Supreme Court justices are on record stating that emerging technologies will likely be the biggest legal challenge of the next few decades (Tolson, 2012). Panelists noted that the judges and legislatures will increasingly face such questions as these:

- Can humans (e.g., witnesses) better identify subjects they see in video feeds, as opposed to or in addition to static mug shots and lineups? How would implicit biases in how witnesses view video be accounted for?
- How can courts address (the often naturally occurring) discrepancies between peoples' statements and testimony on the one hand and video and other sensors on the other?

Many of the CJTFG's legal and policy discussions related to BWCs. Panelists discussed controversies over officer and citizen privacy, such as when a camera should be turned on, and how to deal with cameras not conveying the entire context of an event, such as missing events leading to uses of force. More broadly, panelists described cases of agencies attempting to avoid using BWCs and other monitoring tools to avoid having to take corrective actions; the group discussed whether and how it was possible to incentivize using these tools and deter agencies from avoiding them to dodge accountability.

Increasing commercial pressures on unbreakable encryption are hampering criminal investigations. *Going dark* refers to the trending issue of agencies not being able to get evidentiary data from electronic devices if those devices have strong encryption that source vendors cannot (or will not) override. Recent Apple mobile operating systems, notably, incorporate strong encryption that Apple itself cannot decrypt (see, for example, "Our Approach to Privacy," undated). The CJTFG engaged in substantial discussion on just how much of a challenge this poses to criminal-justice agencies. The group noted that, outside of anecdotal reports about specific investigations being hampered, there are not meaningful metrics regarding the effect that going dark is having on police and other criminal-justice agencies.

The Field Is Getting to True, Criminal-Justice Community–Wide Integration

This theme brings together the overarching demand for criminal justice–wide information integration with a major example of the need for integration (digital-evidence management), combined with a major barrier (have versus have-not agencies).

There are needs for information integration to enable new models of criminal justice across the enterprise. To maximize the potential from emerging technologies that provide more-rapid in-the-field access to information, law-enforcement agencies need to view themselves as one component of a connected criminal-justice enterprise and cycle. The CJTFG discussed the following needs for information integration:

- *improving visibility into offenders:* Data on those under community supervision, as well as coming out of jails and prisons in general, are a major demand for information-led policing models. One panelist noted that community supervision data is the "most pinged" data in the San Diego region's Automated Regional Justice Information System (ARJIS) (ARJIS, undated).

- *improving incident command:* There is a need for information to prevent overloads of communications and personnel during major incidents.
- *predictive analytics to support focused deterrence:* The Cambridge, Massachusetts, police presented on prioritizing the risks of hundreds of thousands of offenders to determine on whom to run focused deterrence efforts.
- *improving entity resolution:* Panelists noted cases of large counties having people do nothing but verify identities for criminal-justice actions (e.g., warrants), driven by the fact that no identifiers go across the criminal-justice system.
- *"one kid, one record":* Systems require a single, secure, high-quality juvenile record per person.[4]
- *police-car telematics for training purposes:* The LAPD reported using in-car telematics to see who needs to improve their driving skills.
- *monitors to protect officers' health:* Panelists suggested that monitors could detect early signs of a heart attack, which is a major cause of death for officers in the line of duty. Panelists did, however, point out a counterrisk, which is what defense attorneys might do with officers' health telemetry data.

This was a central theme because panelists saw information integration as needed to be able to use data in analytics, situational-awareness displays, and so on effectively; this integration is also hampered by a lack of use cases on what types of data integration most need it, awareness of how to do the integration, and surveillance-related policy pressures.

The reality of have and have-not agencies is a barrier to information-sharing and other capabilities across the criminal-justice environment. Budgets, manpower, procurement processes, and other resources vary dramatically throughout the criminal-justice enterprise. Court officers reporting difficulties managing the increasing volumes of BWC footage coming out of police departments was just one example of this that members discussed. Panelists noted a strong need to develop systems and business models that permit "have" agencies to share information and capabilities with have-not agencies in the same region.

There are needs to enable digital-evidence management on a massive scale. Much of the discussion here concerned BWCs footage, but it applies much more broadly to the rapidly emerging family of surveillance-related technologies. There is a broad range of issues that need to be addressed, including

- *formats and standards:* Video and other standards need to work now and in 15 or more years.
- *data redaction to remove faces:* Redacting video is a manually intensive process, with some members reporting their agencies having to redact frame by frame. This is an area being actively researched and developed (e.g., Kanowitz, 2016).

---

[4] The CJTFG did not cover a need for a single, secure, high-quality record per person for adults directly, although, from the tenor of the information-sharing discussions, we can infer that they likely would have concurred with the idea of having one. Other DOJ-sponsored panels have called for having single secure records (for example, Hollywood et al., 2015).

- *data review:* Reviewing video is extremely labor-intensive. This is an active area of federally funded research, on algorithms to flag clips in which something interesting occurs (an actual encounter that turns into a dispute or emergency of some type).
- *storage costs:* Data retention, especially video retention, poses a major challenge to agencies. Future burdens will include DNA collection from crime scenes and suspects.
- *court integration:* There are concerns about prosecutors having neither interoperable systems nor the resources to process large volumes of video data. One solution discussed is to give court attorneys and judges secure log-ins to police video storage systems.

The CJTFG also discussed the role of federal assistance in helping agencies with the data-management challenge. Members asked whether the FBI's Criminal Justice Information Services (CJIS) should provide a federal resource for video footage or whether CJIS should provide security standards and certifications for devices and cloud infrastructure. Panelists noted that CJIS appears to be moving in the latter direction, applying CJIS policy to video footage, starting with video-on-cloud service providers, such as Microsoft Azure and Amazon Web Services.

Safety and Community Relations Need Improvement

This theme connects trends related to societal pressures on guardianship and using technology to improve safety and accountability—most notably, the strong demands for BWCs and improved less-lethal options.

There are increasing pressures to move from "militarized" to "community-based" or "guardianship" models of law enforcement. However, there are also competing pressures to crack down on violence and terrorism—and these pressures can vary enormously from day to day. In the post-Ferguson and -Baltimore environments, demands to demilitarize and "be less violent" appear persistent. Similarly, demands to be more community-centric appear persistent. At the same time, agencies are facing conflicting pressures to crack down on criminal violence, domestic terrorism threats, and organized-crime threats. Penalties for making a mistake in dealing with these pressures can be enormous. As examples, a single video of a police shooting can affect perceptions of police everywhere in the country, while a single mass shooting can drive widespread fears of entire groups of people.

There are increasing pressures for law-enforcement technology in the near future to focus on accountability of all types. These come, in large, part from the trend above, emerging demands to move toward guardianship models of policing (for example, Rahr and Rice, 2015). Panel members noted that, for example, U.S. discussions of BWCs have been almost entirely about holding police accountable. In contrast, discussions of BWCs in the United Kingdom (UK) have been more about filming to gather evidence (and panelists discussed the United States learning from the UK's experience, and vice versa). The assumption is that BWCs and other accountability-monitoring technologies are not going away. Further, panelists noted that there would be legal, political, and civil consequences if these accountability technologies are not present or are mishandled.

Accountability of all types was a major theme of the CJTFG's discussions. These included discussions of privacy protections, ensuring appropriate use of data, documentation of citizen interactions, and improving performance management and officer accountability through data.

There are needs to enable BWC fielding on a large scale. This topic was a focus of the second CJTFG meeting, centered on the LAPD's recent BWC acquisition and associated policies. The LAPD's presentations are substantial issues involving camera fielding, including cost, training, and policy. However, it was also reported that early fielding results are implying that cameras were game changers, with major reductions in uses of force and in assaults on officers observed.

The LAPD's presentation noted that testing candidate camera systems is important before purchase. The agency noted that its vendor selection was based on human-factor considerations reported by officers in the field-testing the different cameras.

Privacy policies with BWCs are a major consideration to be resolved. The issue is complicated by the fact that privacy laws vary substantially by state and put at tension desires to keep sensitive law-enforcement information private (especially footage of bystanders) and desires to make law-enforcement operations transparent. The lack of settled law and case law on surveillance-related technologies is a major barrier here.

Agencies need to further develop and disseminate less-lethal weapons. As background, the panel discussed that "agency heads often feel like they are one YouTube video [of a use-of-force incident] away from a really bad situation in their jurisdictions, which is leading to a great deal of fear facing law enforcement." The panel further discussed that lethal use-of-force incidents generally fell into one of two categories:

- The first includes cases that have *clear interpretations.* These include cases on the extremes—one in which almost all observers would agree that the use of force was justified (e.g., shots against an active shooter) or, alternatively, almost all observers would agree that the use of deadly force was not justified and the responsible officer is sanctioned. Neither of these types of cases usually results in major police–community problems, the panel noted.
- The second includes cases that have *ambiguous interpretations* to different observers. These might include cases in which an officer is technically justified in the use of deadly force, but the use of such force looks like it could or should have been preventable to some observers. An example discussed was a suspect brandishing a knife but not toward anyone in particular. It is cases in this latter category that can result in serious police–community problems: Different interpretations and hence different sides demanding different actions emerge and start conflicting. The panel also noted that this second category exposes the lack of options available to law enforcement short of deadly force to immobilize and restrain a subject posing a threat to the officers and to the public. Members also claimed that there could be a large number of incidents in which, "if a different, non–life taking option was available, it would have been used."

The panel discussed available less-lethal weapons and their limitations, including the current generation of conducted-energy weapons (popularly known as TASERs, after a common former brand name), batons, and pepper sprays. Conducted-energy weapons are seeing increased usage; it was noted that the Chicago Police Department has set out a policy to have at least one per patrol car. Claimed drawbacks, however, included that the weapons' probes often do not penetrate clothes, as well as confusing muscle memory potentially resulting in cases of officers firing guns when they meant to "fire" a gun-like conducted-energy weapon. Batons were described as in general disuse and ineffective. Pepper spray was described as difficult to aim and use to get an immobilizing effect and risked affecting the officers themselves. The result was described as physical grappling, for which officers could lack training, often being the best real-world alternative to using deadly force. As a result, panelists agreed that new development, testing, and fielding of less-lethal devices was needed.

Outside of device development, the CJTFG discussed that the public and policymakers need to be better educated about less-lethal weapons, especially on their limitations. The group noted that agencies (such as Chicago, following the new policy) have held many presentations about these weapons.

New Technologies Bring New Challenges

In general, the implementation of new technology can have serious and unintended consequences—but also major and unanticipated benefits. As noted at multiple points by the CJTFG, technology, whether employed by practitioners or the public, can have a range of unintended consequences. Developments in technology commonly outpace associated developments in law, regulations, policy, culture, and, perhaps most importantly, baseline knowledge of how to use the new technologies effectively in operations. Developing and understanding the key business cases for new technology is one part of addressing this challenge, reducing risk of improper implementations. That said, codifying the major use cases for emerging technology is in tension with a need to let operators develop novel ways to use technology; for example, previously unanticipated uses of information can have major operational benefits.

This was one of the most central themes emerging from the CJTFG. Unanticipated consequences were seen as important to consider for most emerging technology trends.

Touch and rapid-DNA systems that could offer substantial capabilities to law enforcement are emerging, although there are barriers to overcome. Rapid-DNA systems can get DNA typing down to an hour. However, it is very expensive (panelists reported estimated costs of $250,000 for the equipment) with substantial facility requirements.

Touch DNA systems can type DNA with very small samples, such as skin cells that an offender leaves behind after touching an object at a crime scene. It has substantial implications for identifying offenders (Minor, 2013). However, the technology is young, with key algorithms still under development.

Finally, the new technology of DNA phenotyping can create an image of what a subject looks like based on the subject's DNA (Pollack, 2015). This technology is very early in its development as well.

Panelists noted that, in addition to costs, DNA technologies would require large amounts of data handling to manage all of the typing data. Further, panelists noted that DNA technologies will raise substantial privacy issues and legal challenges—notably, on conditions (which vary by state) under which a person's DNA can be sampled and which typing databases are subject to search and when. Again, this is an example of the broader emergence of a family of common policy challenges around emerging surveillance technologies.

Technologies to detect firearms and other weapons remotely might be emerging. Panelists noted pilot efforts to use millimeter-wave technology to detect weapons from standoff distances, although they noted that examples to date appear to be van-sized devices with uncertain accuracies (see, for example, Andrews et al., 2013) Panelists also noted significant challenges and variations in state and local law in the conditions under which such a device might be used.

Recommendations and Actions

As described above, we analyzed the notes from CJTFG meetings 1 through 3 to identify recommendations providing specific ways ahead to address the trends and themes. The full CJTFG reviewed and approved the recommendations at the fourth meeting. BJA further identified three short-term action items from the CJTFG's discussions, and the CJTFG reviewed and approved an additional resolution on data exportability at the fourth meeting. Below, we present the recommendations and actions items, by the theme they support.

Findings to Address the Lack of Business Cases and Processes

The lack of business cases and processes describing both best practices for using technologies and best practices for implementing them were identified as being major barriers to criminal-justice adaptations to new technologies across the board, as well as being behind major security, privacy, and civil-rights risks. Figure 1.4 summarizes this theme, constituent trends, and ways ahead, using a small mind map.

Action 1: Start developing core business cases with information exchanges to external service providers.

Perhaps the top callout of the CJTFG meetings was whether we can capture the most-promising applications and corresponding requirements for using data and analytics in business cases. As a start, BJA chartered two Global Justice Information Sharing Initiative task teams, jointly sponsored with the CJTFG, to identify cases for exchanging critical information with organizations outside of the traditional criminal-justice system in pursuit of criminal-justice objectives.

One team is considering two-way exchanges between medical and justice organizations (see, e.g., Justice to Health). Building on earlier work characterizing needed information exchanges in this area in general (Global Strategic Solutions Working Group, 2014; Global Standards Council's Justice-to-Health Services Task Team, 2014), this group is identifying strategies on key information exchanges needed in support of continuity of care during reentry,

Figure 1.4
Responding to the Lack of Business Cases and Processes

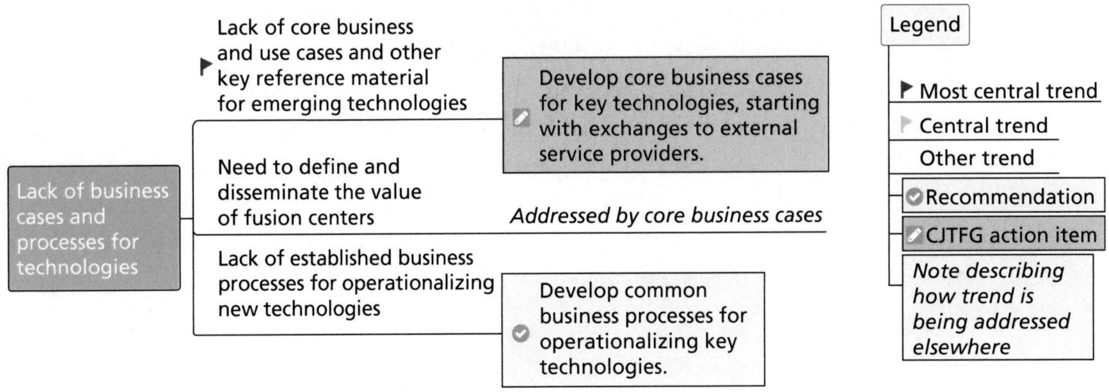

RAND *RR1987-1.4*

countering the "school-to-prison pipeline," and providing care in response to the U.S. heroin and opiate epidemic.

The second team is providing a broad coverage of two-way "nontraditional" exchanges needed between justice agencies and external agencies in support of criminal-justice reforms intended to reduce both incarceration and recidivism. These include pretrial reform, diversion of offenders away from incarceration (such as drug courts, problem-solving courts, and Justice Reinvestment Initiative programs), and smart reentry of those coming out of incarceration. External agencies mentioned include government identity (social security or department of motor vehicles), education, housing, welfare, mental health, social services, and medical (including coordination with the Justice to Health team).

The charters for both task teams are found in Appendix E, available online.

An early finding is the need for centralized case management. The task teams' work to date has identified substantial commonalities across many of the business cases, especially those related to initiatives to reduce incarceration and recidivism (whether focused on pretrial, diversion, or reentry). These include the needs for a *common case-management function* to "wrap clients with services," holistically meeting their needs in a way to maximize their chances of success by avoiding recidivism and becoming productive members of society. This case-management function needs to oversee and direct a series of information exchanges to get those services performed successfully, including risk- and need-assessment data, criminal-justice history data, medical data, mental-health data, and substance-abuse treatment data (with prescription data cutting across all three). Further, in directing these exchanges, the case-management function needs to navigate the client's consent to share the needed information, because of the Health Insurance Portability and Accountability Act of 1996 (HIPAA) (Public Law 104-191) and related requirements. Figure 1.5 diagrams what this broad need for case managers to direct services looks like.

Recommendation 1: Develop common process templates for operationalizing new technologies.

In addition to core business cases and supporting specifications for how new technologies should be employed, additional components that should be included in the common process templates include

- total cost assessments across the complete life cycle
- integration and interoperability requirements
- requirements for community and external expert participation
- cyber, civil, and privacy rights protections, along with other policy provisions, building on the IACP Technology Policy Framework (IACP, 2014). Recommendation 3 provides additional information on incorporating these protections into common processes.
- other governance implications
- change-management requirements and implications. Panelists noted the Boston Consulting Group's duration, integrity, commitment, and effort formula (Sirkin, Keenan, and

Figure 1.5
An Emerging Need for Comprehensive Case Management

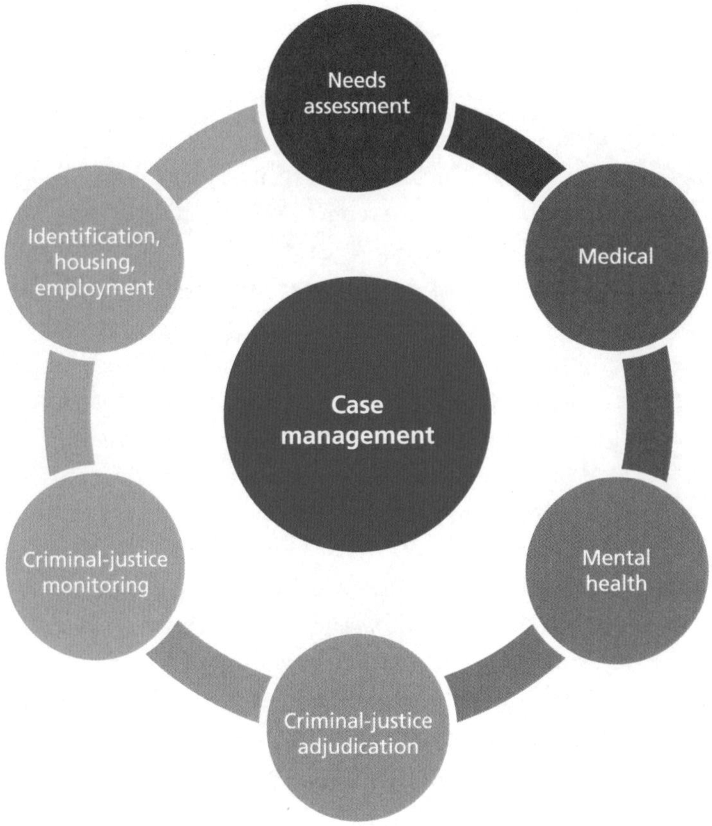

RAND *RR1987-1.5*

Jackson, 2005), which weighs the probability of success of a project (like fielding a new technology) by using the following factors:

- duration, with shorter milestone cycle times being lower risk
- integrity (actually performance capability) of the project team
- commitment to the project from both senior leadership and the intended users
- effort required to implement the new technology.

Findings Regarding the Emergence of Big Data, Analytics, and Associated Challenges

These three ways ahead, summarized in Figure 1.6, address the overarching theme of the emergence of big data and analytics and the challenges of using them.

Recommendation 2: Conduct research to improve how criminal-justice technology information is made available to both practitioners and researchers.

Figure 1.6
Responding to the Emergence of Big Data, Analytics, and Associated Challenges

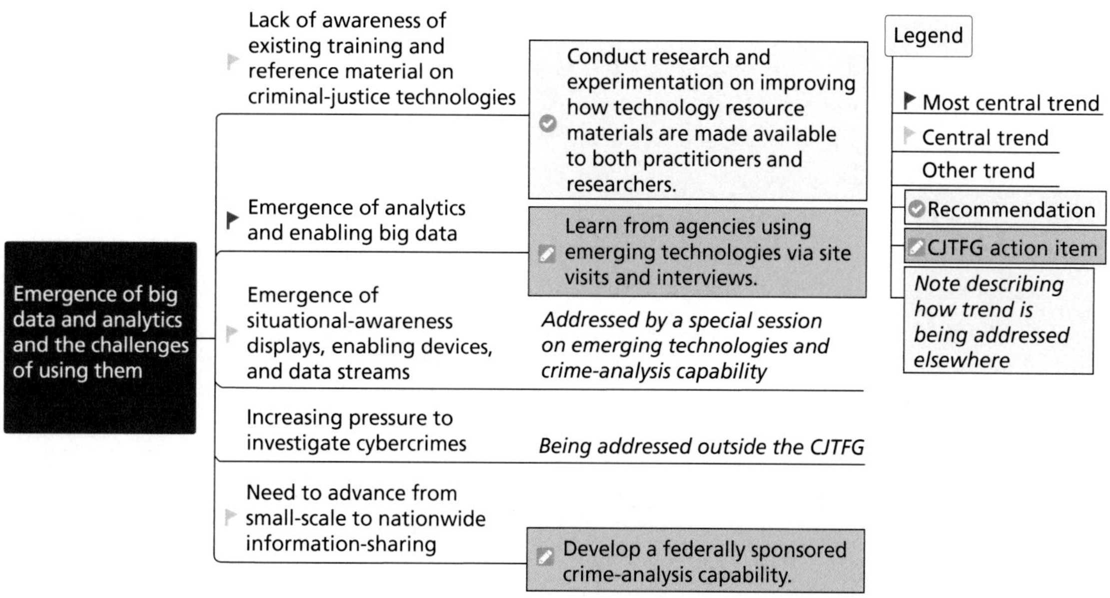

RAND *RR1987-1.6*

As discussed earlier, many technology references are available but are widely unknown and often not well suited operationally to criminal-justice practice practitioners. As a few exceptions, panelists noted that they knew of agencies learning about technologies from the IACP's *Police Chief* magazine and the Police Executive Research Forum's technology articles; the panelists did not mention DOJ-sponsored resources as being commonly used.

Further, panelists noted that there are areas of technology research and development about which few in criminal-justice practice know but are becoming widespread many other fields. Machine learning is a key example (with predictive policing and risk-assessment instruments being the partial exceptions to the lack of knowledge about it), but some biology, medical, and physical science breakthroughs are also not getting into criminal justice.

This recommendation calls for conducting research and experimentation on improving how technology resource materials are organized within and across major reference sites, how technology material is presented, and how new technology resources are advertised to the practitioner community. The panel suggested that researchers learn from mechanisms that other industries use to learn about and incorporate new technologies, with agricultural extension programs given as one example.

Action 2: Learn from agencies using emerging data science technologies.

The CJTFG is recommending that BJA sponsor site visits and interviews to agencies using emerging technologies to assess what might be done with emerging data science technologies that would be of most value to criminal justice. The emergence of big data and analytics is one of the most central issues to accelerating criminal-justice use of technology. The focus should be on visiting agencies using technologies today that are likely to be ubiquitous in two to five years (e.g., the equivalent of today's smartphones and fitness trackers). Questions to be asked might include

- What can agencies do with big data from multiple sources (including video and IoT sensors) that would be useful for criminal justice? At a lower level, what are the impacts, requirements, and opportunities from increasing migration to cloud and shared-service models for IT?
- What do agencies (and their research and commercial partners) find to be technologies that are likely to make a big operational impact over the next few years? Assessment questions can look for technologies that are becoming ubiquitous (e.g., everyone now carries a smartphone), have high political pressure for use, work well enough and are affordable enough to support radically changing a core business process, or require major changes to hiring criteria and training protocols.

Action 3: Develop a concept for a federally sponsored crime-analysis capability.

CJTFG members discussed federal sponsorship of a core crime-analysis capability that would use federal databases as data sources. This model would provide incentives for agencies to contribute data to these databases.

The group discussed leveraging new, open-source cloud and information integration tools to provide a simple-to-use environment (multistandard, multijurisdiction) that can provide many of the key crime-analysis outputs very easily. The first tool would be crime mapping for both police and the public. The principal incident input would be National Incident Based Reporting System (NIBRS), which has the benefit of providing a strong incentive for agencies to migrate their federal crime reporting to NIBRS from legacy Uniform Crime Reporting (UCR). Similarly, the same resource could include uploads of National Data Exchange (N-DEx), Law Enforcement Information Exchange (LInX), and COPLINK data, which provide information about known people, organizations, and assets of interest. Together, the two types of data can drive key crime analyses. From a design perspective, the environment could be set up as a mash-up of existing core open-source tools (e.g., R, OpenStreetMap, Python) and the government-owned systems (N-DEx and other CJIS systems). Following discussion of the idea, BJA tasked the CJTFG to develop a brief white paper describing the proposed capability, which we present as Appendix C.

Findings to Address Security, Privacy, and Civil-Rights Challenges

Panelists noted that agencies implementing technologies without advance policy work, or "arrogance" that ignores public concerns about surveillance, can incur substantial risks. In an extreme case, agencies can end up using systems in politically unacceptable ways that create community-relations crises and can even put all of criminal justice's access to a technology at risk. This set of recommendations, summarized in Figure 1.7, addresses security, privacy, and civil-rights challenges to using new technologies effectively and with community support.

Figure 1.7
Responding to Security, Privacy, and Civil-Rights Challenges

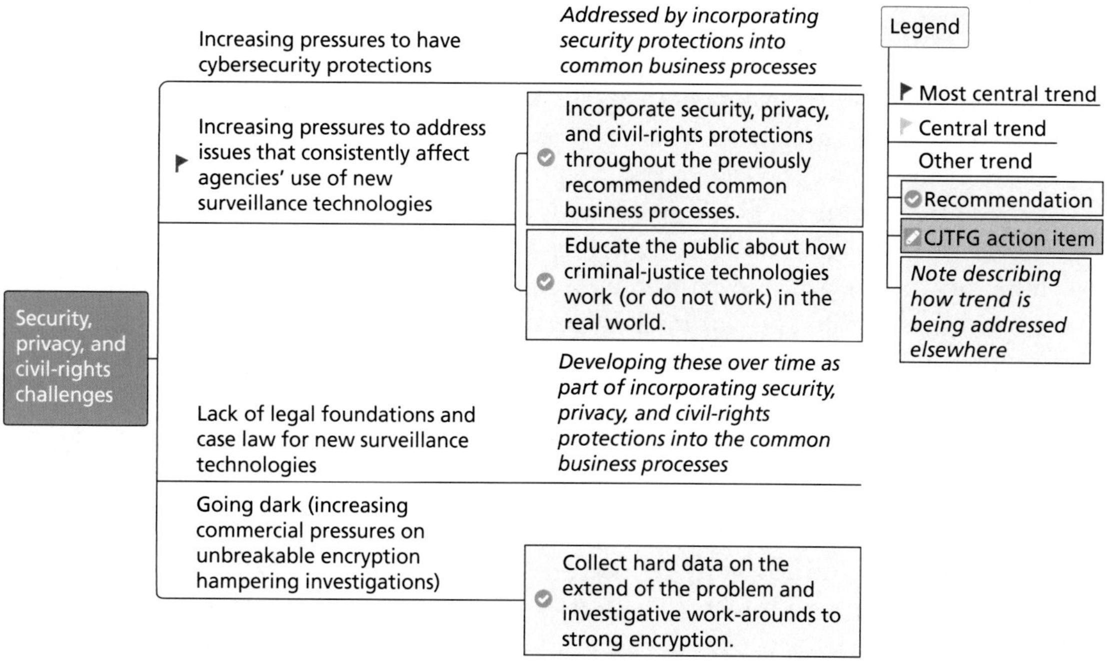

RAND *RR1987-1.7*

Recommendation 3: Integrate security, privacy, and civil-rights protections into the common business processes (from recommendation 1) for adapting new technologies.

The goal should be to incorporate these protections throughout the development, acquisition, and operational life cycle. Key aspects of how to do this include both (1) building off of prior work, providing agencies with general templates and checklists and specific case studies, and (2) providing guidance on how to involve communities and local policymakers in genuinely open processes in setting policies and procedures appropriate to the state and locality.

For (1), panelists noted that many agencies do not have resources to engage in comprehensive policy work, which, in turn, can lead to agencies just adopting technologies, leading to potential community-relations crises as a result. Thus, prior blueprints from which to work are needed. These should start with clear descriptions of how the technology actually works and clear bounding conditions on both how it could be used and how it should not be used. Such descriptions should be available from the business cases for new technologies. These should include checklists on how to consider security impacts, privacy impact, and demographic and socioeconomic impacts of proposed uses.

Panelists also noted that agencies need guidance on what not to do: One example was taking public stances that are counterproductive, such as being against transparency (such as not allowing recorded people to see the video of themselves); a second was permitting mission creep, with new technologies used in ways far beyond what was initially intended or publicized. Panelists noted that Georgetown Law's recent report on facial recognition (Garvie,

Bedoya, and Frankle, 2016) shows what can happen when technology uses run ahead of policy frameworks: The report claimed that more than half of U.S. adults were in law-enforcement facial-recognition networks, with few systematic protections on how those networks are used.

As example efforts to emulate, panelists noted that the U.S. Department of Homeland Security (DHS) did a good job in its policy review of LPRs, which included direct participation of privacy experts and organizations (DHS, 2015). The ongoing IACP/IJIS Institute image technology discussions have brought together industry and practitioners to help define "appropriate use" and develop policies to ensure appropriate use of new technologies.

For (2), panelists noted that it was not possible to create one-size-fits-all technology policies because state and local laws, as well as local operating contexts, are so different. Thus, community involvement is needed to customize policies suitable for each jurisdiction. Panelists noted that civil and privacy rights advocates have generally been disappointed in the lack of community engagement and transparency involved in setting technology-usage policies to date (or, as noted, the lack thereof). Guidance needs to show how community-engagement processes can be done at fairly low cost and effort, given that most agencies will have very limited resources.

Recommendation 4: Educate the public on how criminal-justice technologies work or (do not work) in the real world.

Panelists noted two aspects of this. The first is developing informative approaches to reduce the public's false expectations of what technologies criminal-justice agencies have and what they can actually do (known to the panel as reducing the "CSI effect").[5] The second is, to alleviate fears of those technologies being abused, developing informative approaches to explain how technologies are being used in practice.

One initial approach here would be to develop straightforward lists of what particular technology applications do and do not do and what criminal-justice agencies are and are not doing with them; this approach is readily actionable and could be a good gateway to longer conversations with community stakeholders. In the latter case, such lists could, over time, become lists of best practices for criminal-justice agencies.

Recommendation 5: Collect data on the going-dark problem.

There has been a substantial debate on how much of a challenge this actually poses to criminal-justice agencies outside of anecdotal reports about specific investigations being hampered. There is a need to collect hard, quantitative data on the extent of the problem—specifically, to count investigations in which (1) it was strongly believed that critical evidence needed to proceed with the investigation was on an encrypted device and (2) the investigation was stymied as a result of being unable to decrypt the device. This leaves aside cases in which the evidence was seen as useful but not really necessary and cases in which investigators

[5] Note that the real-world impact of unrealistic technology expectations is difficult to assess; as an example, the CSI effect has not made statistically significant differences in how juries arrive at verdicts (see Goodman-Delahunty and Verbrugge, 2010, for a review of relevant research).

thought something useful might have been on the device. It will also be important to identify work-arounds for investigations if or when strong encryption becomes ubiquitous.

Findings on Getting to Field-Wide Integration

These recommendations and actions, summarized in Figure 1.8, seek to help attain information integration throughout the criminal-justice community.

Recommendation 6: Research changing cultures to support information-sharing and safeguarding.

The need to change the culture of organizations is one that is commonly presented as a major barrier to improved information-sharing and safeguarding. Rather than treat culture as an insurmountable barrier, here we call for research on methods to help change cultures to better support information-sharing and safeguarding. CJTFG members suggested examining models from other fields to help do this. One example was social-network analysis to identify emerging thought leaders and foster peer-to-peer relationships. Members also felt that there was a "need to go beyond" current information-sharing forums, such as the IACP's *Police Chief* magazine, mentioned earlier as a solution to law enforcement's perceived lack of engagement with technological advances.

Recommendation 7: Develop regional models for information-sharing capabilities.

Figure 1.8
Ways Ahead to Support Getting to Field-Wide Integration

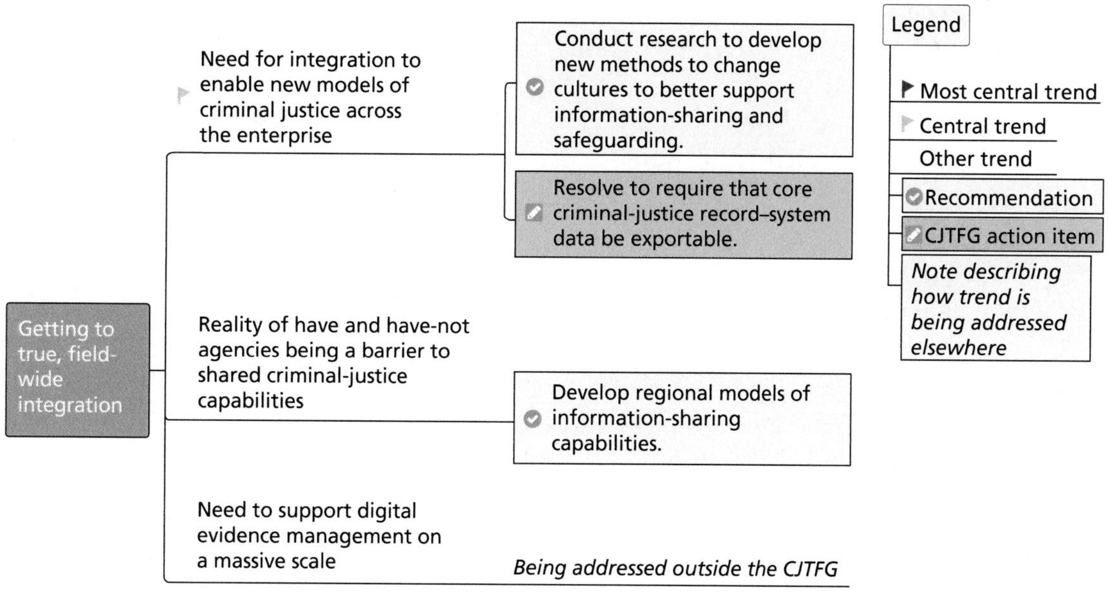

RAND *RR1987-1.8*

This recommendation would identify and disseminate examples of regional models for sharing capabilities, then develop promising practices based on these examples. It is specifically intended to help ameliorate the have-and-have-not problems within a region. Types of examples would include membership cooperatives, in which a lead agency purchases an IT capability or service and other agencies contribute subscription fees to get access to that same service. It also includes statewide and regional models in which a large organization creates an online information resource (e.g., a regional data-exchange warehouse) that serves all agencies within a state or region. Example types of information that might be shared regionally, based on CJTFG discussions, include

- life-critical data, which are data needed to positively identify a person (potentially including biometrics), specify any high risks to personnel from a person (or people who might be at a location), and specify any enforcement actions needed (i.e., carry out an arrest warrant or notify a community corrections officer)
- corrections data, especially on those under community supervision, those coming out of incarceration, and relevant points of contact in community corrections agencies
- data needed to support regional incident responses.

Action Item 4: Require that core criminal-justice record–system data be exportable.

The CJTFG endorsed a Global Advisory Council resolution to require that core criminal-justice record data be made *fully exportable.*

As background, SEARCH, which is a national consortium on interjurisdiction justice information–sharing, has identified some cases in which providers of core record systems (law-enforcement record-management systems, CAD systems, case-management systems, jail-management systems) simply would not or could not implement federally sponsored data-sharing standards in their systems directly. Recent advances have made it much easier to create interfaces that take data stored in one format and rapidly convert them to other standards (via open-source tools, such as Apache Camel, for example) for wide-scale sharing, effectively dealing with this problem. However, SEARCH further found cases of systems not allowing agencies to export data, as well as software licenses stating that the vendor owns the agencies' data and preventing them from building interfaces to export their data out of the system. Global is developing guidance to help agencies avoid such pitfalls when they purchase new systems.

The CJTFG endorsed a resolution that will be the core part of a larger guidance package. It will require agencies to have unfettered access to use their own data, as follows:

- Vendors must give agencies a license to use and export their own data, as they see fit, for any purpose they wish.
- Vendors must have a feature allowing agencies to export system data in a format that they can readily translate.

Once these provisions are in place, an agency can take advantage of many increasingly low-cost solutions that translate data to standards, such as the NIBRS standard (for crime events) and the national N-DEx standard (for data about people, places, and assets).

CJTFG members strongly endorsed this resolution, noting that, "if this idea had existed a decade ago, [information] searches and the transitions to NIBRS and N-DEx would be an order of magnitude better" and that the "the issue isn't agency willingness to participate in information-sharing—they are very willing—it's figuring out interfaces and how to pay for them." The complex text of the resolution is in Appendix F, available online.

Note on Work Outside the CJTFG on Digital-Evidence Management

The panel discussed CJIS moving in the direction of providing security standards and certification for camera devices and cloud infrastructure. The CJTFG discussed that CJIS is probably going to apply CJIS policy and certifications to BWC video, starting with procedures for certifying video on cloud service providers.

The group also discussed the idea of CJIS setting up and maintaining a federally supported cloud for video that would be open to state and local agencies; the major limitation here is that different states and localities have very different laws, policies, and traditions on whether and how such a cloud could be used, probably creating a national storage capability infeasible at this time.

Findings on Improving Safety and Community Relations

This set of recommendations, summarized in Figure 1.9, concerns addressing societal pressures on guardianship and using technology to improve safety and accountability—most notably, the strong demands for BWCs and improved less-lethal options.

Figure 1.9
Ways Ahead to Improve Safety and Community Relations

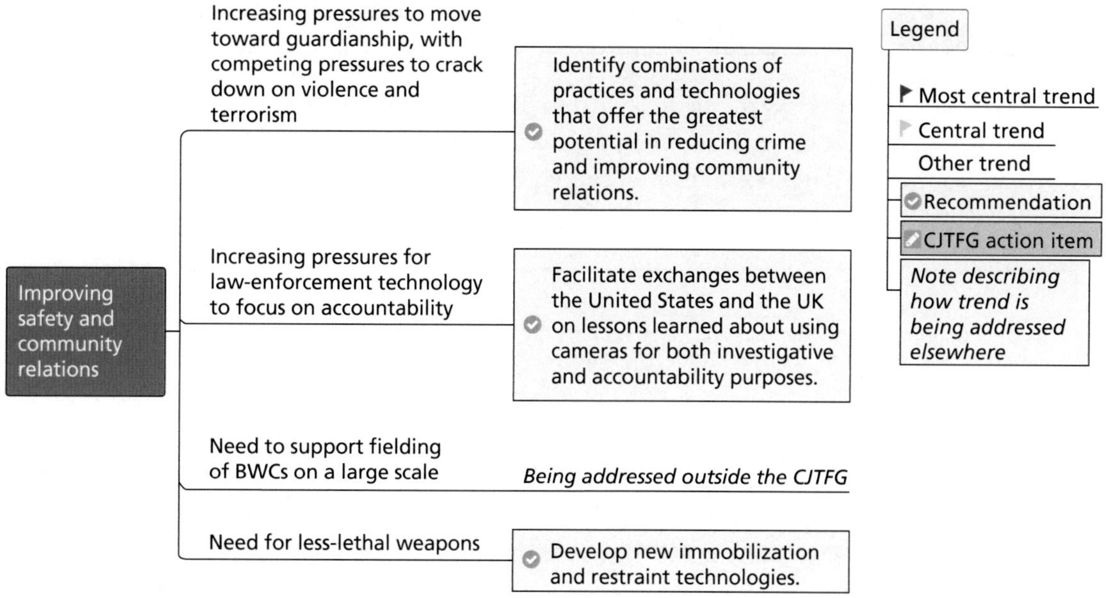

RAND *RR1987-1.9*

Recommendation 8. Identify practices and technologies that can both reduce crime and improve community relations.

This recommendation calls for research and evaluation to find those combinations of criminal-justice practices and supporting technologies that offer the greatest potential for effectiveness in reducing crime and terrorism and in improving community relations. Specific questions to be addressed include

- *guardianship policing:* How does law enforcement deal effectively with heavy public and political pressures to migrate away from "militarization" and toward "guardianship" (referred to, for example, as "the post–post-9/11 era")?
- *conversely (to guardianship policing):* At the same time, how does law enforcement deal with rapidly shifting and competing pressures to crack down on radical violence and violent crime? In the modern media and Internet environment, all of these pressures can flare up rapidly in response to major incidents—sometimes at the same time (e.g., stories on the San Bernardino shootings versus a video release of an officer-involved shooting in Chicago).

Given the large number of articles and data to date on evaluating various policing strategies, the short-term focus can be on reviewing and analyzing prior research to find the needed combinations.[6]

Recommendation 9: Explore international exchanges on using cameras for investigative and accountability purposes.

As noted, in the UK, the primary driver of camera use is to gather evidence, whereas, in the United States, the driver is to hold police accountable. It appears that both countries could learn a great deal through exchanges about lessons learned on using BWCs for both investigative (evidence-collecting) and accountability purposes.

Note on Outside Work on Supporting Large-Scale BWCs

The CJTFG emphasized the importance of supporting fielding of BWCs. At the same time, a great deal of work is going on in this area, with federal grants for cameras and BJA's Body-Worn Camera Toolkit being just two examples.

Recommendation 10: Develop new immobilization and restraint devices to provide alternatives to lethal uses of force.

---

[6] RAND is conducting an internally funded research project very close to this recommendation; it is reviewing prior evaluation studies to identify policing strategies that appear most effective and equitable.

The CJTFG noted that improved immobilization and restraint devices are needed—ideally, devices that can perform both functions at the same time, with low risk to all involved parties.

As noted earlier, the CJTFG discussed limitations of current less-lethal weapons, with one point of discussion being that "the best real alternative [to deadly force] is poorly or under-trained officers grappling with suspects." Panelists went on to suggest developing devices that might focus on immobilizing legs and arms, with one noting that "we haven't really used science to understand the dynamics of an arrest situation and develop options."

The CJTFG discussed elements that could inform development and testing of new less-lethal devices:

- Inventory currently available law-enforcement, military, and other nonlethal equipment.
- Inventory past and ongoing efforts to develop nonlethal options, including failed efforts.
- Develop a concise set of scenarios in which law enforcement would wish to use a nonlethal option (e.g., stop a fleeing suspect, subdue a suspect with a knife or nonshooting weapon) and assess the specific needs in those scenarios (e.g., immobilize legs, render a suspect unable to hold or use a weapon).
- Identify conditions and requirements to prevent new weapons from being seen as making policing difficult or dangerous.
- Identify appropriate safety, privacy, and civil-rights policies and procedures.
- Get feedback from a broad range of law enforcement and other stakeholders throughout.

Findings on Addressing New Technologies and New Consequences

This final recommendation, summarized in Figure 1.10, addresses the theme of addressing new technologies and new consequences.

Recommendation 11: Assess the potential of remote weapon-detection capabilities.

Figure 1.10
Responding to New Technologies and New Consequences

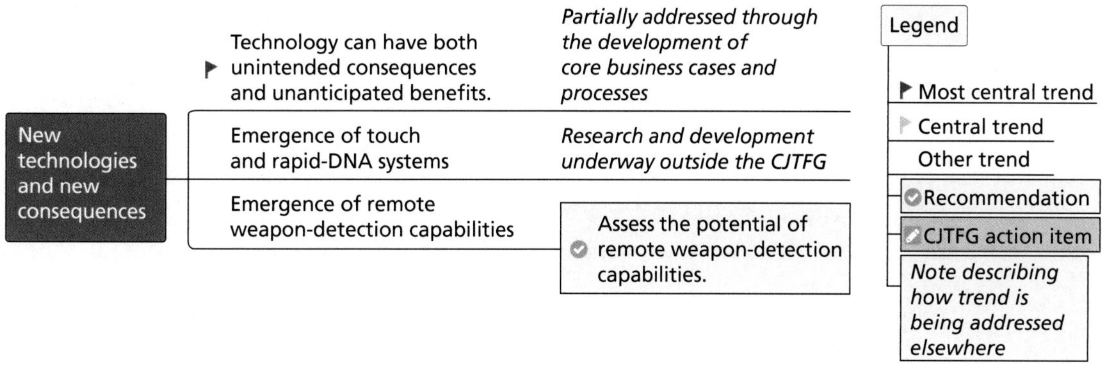

RAND *RR1987-1.10*

Panelists noted that the ability to detect a weapon from a distance, especially a firearm, could be enormously beneficial to law enforcement, community corrections officers, and facility security. That said, there was also skepticism about the size, power, accuracy, and cost of such devices. It did appear to be worth assessing the state of the art in this area. If the technology appears promising for adapting into deployable devices, it will be valuable to look at the circumstances under which these devices should be used from legal and operational perspectives, as well as needed safety, privacy, and civil-rights protections.

Note on the Emergence of Touch and Rapid-DNA Systems

This was another technology that the panel thought might have very substantial benefits for the criminal-justice system but seen as one that was both maturing and had substantial research already being done on it.

Note on the General Trend of Technology Having Unintended Consequences and Benefits

This was seen as an underlying fact of which practitioners and developers need to be aware, rather than a problem with a concrete solution. The CJTFG did discuss that this trend can be partially ameliorated through developing core business cases and processes for technology adoption.

Conclusions

BJA charged the CJTFG to assess major trends likely to affect the criminal-justice system in the next three to five years and identify ways ahead to address the problems and opportunities they pose. In response, the CJTFG has covered a very wide range in topics in its meetings, identifying close to two dozen trends, assessing their impacts, and developing more than a dozen ways ahead in response. Further, despite this wide range, the CJTFG did find something of an overall narrative linking many of the trends: Most of the opportunities and developments are related to information processing and analysis, but taking advantage of information-processing advances successfully is greatly hampered by a lack of capability on business cases, implementing processes, and security, privacy, and civil-rights protections. Getting to true, field-spanning integration across agencies' systems is especially challenged by these issues and further challenged by "built-in" noninteroperability in some widely used systems. These issues also challenge being able to have technologies (mostly IT) contribute successfully to improving safety and community relations and take advantage of other new technologies (that themselves usually have major data components).

The CJTFG's responses are coordinated as well. The CJTFG helped to move ahead on business-case development; cosponsoring teams developing "nontraditional" criminal-justice exchanges; developing the concept (core case) for a federally sponsored, nationwide crime-analysis capability; and resolving to fix a major barrier to information-sharing—licenses and software written to preclude it. It has called for development of an integrated but customizable business process for implementing new technologies that incorporates provisions on security, privacy, and civil-rights protections into the process (technical and policy) rather than them being afterthoughts. To overcome the lack of knowledge of technologies, process, and security and civil-rights protections, it is calling for research on improving how educational materials are distributed to practitioners; to overcome cultural barriers to the same, it is calling for research on how to best influence criminal-justice organizational cultures. To address the pressing challenges of improving safety and community relations, it is recommending research both on improved strategies and on improved equipment (immobilization and restraint devices).

Recall that Table S.1 in the summary provided an overview of the top themes, trends, and ways ahead that the CJTFG identified. All are intended to help the criminal-justice field deal with the opportunities and challenges posed by today's ongoing technology and social trends.

APPENDIX A

List of Criminal Justice Technology Forecasting Group Members

In this appendix, we list the positions that members held when they served on the CJTFG.

- Peggy Bell, executive director, Delaware Criminal Justice Information System
- Jean Bousquet, director and chief information officer, Wisconsin State Courts Office
- Ronald Brothers, chief information officer, Information Technology and Communications Division, Maryland Department of Transportation
- Scott Came, executive director, SEARCH Group
- Thomas Clarke, vice president, research and technology services, National Center for State Courts
- Mitch Cunningham, deputy chief, Investigative Bureau, Wilmington (N.C.) Police Department
- Phillip Goff, president, Center for Policing Equity, John Jay College of Criminal Justice
- Maggie Goodrich, chief information officer, Los Angeles Police Department
- John Hollywood, Ph.D., director, Information and Geospatial Technologies Center of Excellence, National Law Enforcement and Corrections Technology Center
- Jonathan Lewin, commander and managing deputy director, Chicago Office of Emergency Management and Communications, Chicago Police Department
- Cherie Lingelbach, policy and implementation manager, juvenile justice information system, Oregon Youth Authority
- Mike Overton, chief, Information Services Division, Nebraska Crime Commission
- Jerry Ratcliffe, Ph.D., director, Center for Security and Crime Science, Department of Criminal Justice, Temple University
- David Roberts, senior program manager, programs, International Association of Chiefs of Police
- Anne Roest, commissioner, New York City Department of Information Technology and Telecommunications
- Donna Roy, executive director, Information Sharing Environment Office, U.S. Department of Homeland Security
- Cynthia Rudin, associate professor, Sloan School of Management, Massachusetts Institute of Technology
- Pam Scanlon, executive director, Automated Regional Justice Information System
- Teresa Takai, chief information officer, U.S. Department of Defense
- Sean Thakkar, executive director, Connecticut Criminal Justice Information System Governing Board
- Joseph Wassel, U.S. Department of Defense
- Harlan Yu, principal, Upturn.

APPENDIX B

Technical Appendix

The full process and methodologies used to conduct the CJTFG's meetings and analyze the results are provided in this appendix.

Meetings 1 (2014) and 2 (2015) featured a combination of guest speakers on technology and criminal-justice topics and focused breakout groups. Meeting 1, held in March 2014, included panels and presentations on current technology initiatives, including those highlighting initiatives by the National Institute of Justice, Global Advisory Committee, BJA, IACP Technology Center, Regional Information Sharing Systems, and the program manager for the Information Sharing Environment's initiatives on enabling the future of information-sharing. Panelists also discussed and rank-ordered six technology areas by how much impact they would likely have on criminal justice in the next three to five years (i.e., out to 2020). These technologies were, in order of "most expected impact" to "least impact,"

- mobile technology
- privacy and civil-liberty issues
- system interconnectivity
- biometrics and facial recognition
- big data
- cloud computing.

Meeting 2's breakout topics included the operationalization of mobile technologies; sharing court and corrections data with law enforcement; real-time crime centers and fusion centers; and "removing offender anonymity to better deploy resources by overlaying data from warrants, probation/parole, juvenile justice, and social services." The meeting also saw special presentations on the future of BWC deployment by the LAPD (which hosted the meeting) and on preventing and responding to cyberattacks from the program manager for the Information Sharing Environment.

Following meetings 1 and 2, we reviewed the notes from these meetings to identify an initial set of high-level trends and impacts. These were

- information-sharing and safeguarding
  - need for business or use cases for key technologies
  - limited progress to date on information-sharing outside of small-scale efforts
  - widespread privacy and civil-rights concerns about information-sharing technologies
  - substantial efforts to develop cybersecurity infrastructure and resources
  - limited progress on cyber outside of small-scale initiatives

- video and other sensors
  - technology, such as BWCs, being a driver of accountability for both practitioners and the public
  - human-factor issues being major determinants in which BWCs have been selected or found to be most useful
  - vehicle telematics to track officers' driving and location
  - growing interest in using sensors to monitor officers' health
  - widespread privacy and civil-rights concerns about sensors (surveillance) technologies
- digital forensics
  - limited court capabilities to access and manage large amounts of digital evidence
  - growing consideration of laws and policies that trade off privacy and public access
  - strong demand for, and development of, technologies to automatically redact video feeds
- operational technologies (vehicles, gear, weapons, facilities—i.e., non-IT)
  - affordability and accessibility of new technologies for perpetually budget-challenged criminal-justice agencies.

Meeting 3 (2016) was a structured discussion to further develop the trends and their likely impacts on criminal justice, as well as potential responses to the trends. To develop the list of trends, we asked the group to consider technology and related social and political trends. We asked members to focus on trends expected to have the greatest impact on criminal justice in the next three to five years. We also asked them to consider the effects of interactions between trends; we provided, as an example, the fact that the combination of inexpensive video cameras, putting those cameras on widely distributed smartphones, and being able to upload that video to social media sites for public consumption enabled an explosion of the public's filming of law-enforcement activities.

We asked panelists to assess the impacts on criminal-justice operations in the next two to five years, including both opportunities and threats. We also asked them to consider possible measures for assessing impact and, if possible, develop a high-level consensus on the likely magnitude and breadth of the potential impact on criminal justice. Key dimensions along which impacts could occur included

- people: e.g., training, recruiting, retention
- process: tactics, strategies, and procedures
- organization: agency structures and budget
- technology: tools, equipment, and acquisition.

To develop potential responses, we asked group members to consider ways to advise agencies on addressing trends more effectively and ways to advise developers and funders on actions that would make technologies more useful—or less of a threat. We asked them to identify any known evidence on how potential solutions have worked to date (i.e., results of field experiments), or at least logical requirements for success, and, if possible, develop a high-level consensus on the potential value and feasibility of proposed responses.

The group's discussions were subdivided by topic, using the initial list of trends described above (i.e., we hosted discussions on information-sharing and safeguarding, video and other sensors, digital forensics, operational technologies, and "anything not previously covered").

Each topical discussion began with brief "fire-starter" discussions to provide examples of new technological and educational developments; these included presentations on the IACP's Law Enforcement Cyber Center (Law Enforcement Cyber Center, undated [a]); the OpenSource-Policing resource portal; predictive policing of those identified as being of high risk of being a party to violence in Chicago; BJA's Body-Worn Camera Toolkit resource portal (BJA, undated); findings from the "Data and Civil Rights: A New Era of Policing and Justice" conference in 2015 (Data and Civil Rights, 2015); and the graduated-reentry concept for reducing incarceration rates.

We captured the major points of the CJTFG's deliberations as they happened using mind mapping, via the software package XMind. A mind map is a diagram that captures information in a hierarchical format that looks like an expanding tree, showing how that information is organized into varying levels of groups (e.g., branches, leaves), as well as how different pieces of information relate to one another (see, for example, Pinola, 2013). Our approach to capturing the deliberations was based on the Futures Wheel (Glenn, 1972), as shown in Figure B.1. The idea is to capture an initial trend, potential direct (first-order) impacts from that trend, potential subsequent indirect (second-order) impacts following from the first-order effects, and the links between them, in an expanding radial diagram.

We expanded the Futures Wheel concept to capture trends, trend details, impacts (opportunities and threats), potential responses, and key relationships between them. Figure B.2 diagrams how we implemented the expanded concept as a mind map in XMind.

At the end of meeting 3, we identified 22 trends and associated impacts, potential responses, and relationships.

Figure B.1
The Futures Wheel

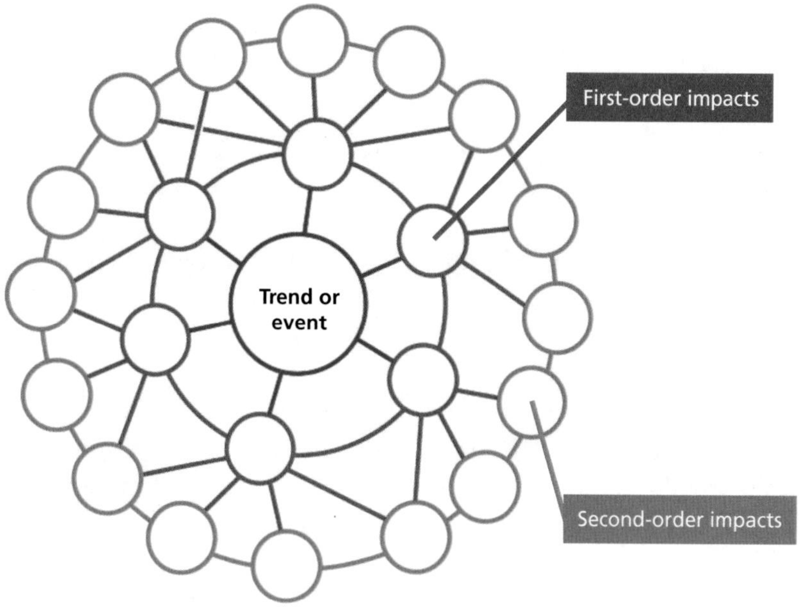

SOURCE: Zapyon, 2009. Creative Commons Attribution 2.5 Generic license.
NOTE: The blue center is the first-order effect. Red indicates a second-order effect, and so on.
RAND *RR1987-B.1*

Figure B.2
Implementing an Expanded Futures Wheel as a Mind Map

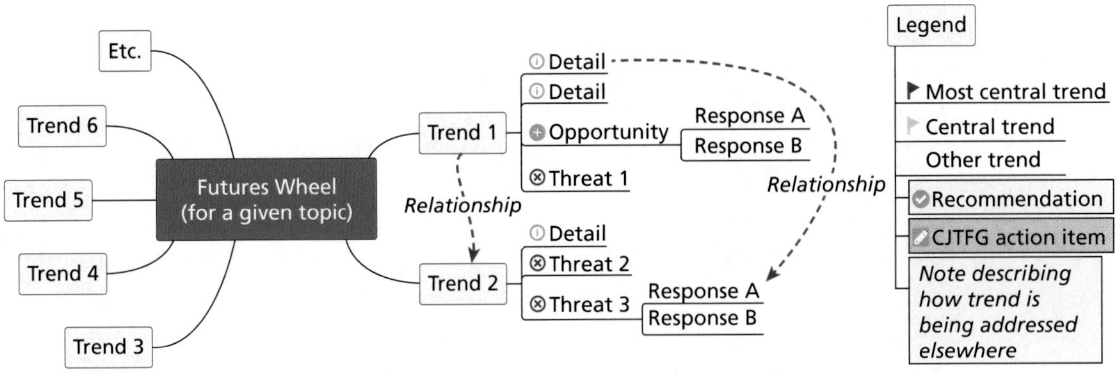

RAND *RR1987-B.2*

Analyzing the Trends, Impacts, and Responses, Post–Meeting 3

Following the meeting, we reviewed the complete mind map of trends generated during the meeting, along with notes taken about them, to identify relationships between the trends. These relationships address

- *key dependencies between trends.* For example, the lack of establishing business processes for implementing technologies is driven in part by the lack of use cases.
- *cases in which one trend is, in part, an example of a more general trend.* For example, the need for information-sharing use cases is one part of the larger trend on information integration to enable new models of criminal justice.
- *cases in which one trend is particularly affecting another trend.* For example, common pressures to address certain policy concerns (privacy, civil rights, and information assurance) about surveillance-related technologies in general are affecting the surveillance technology– and analytics-related trends.

Table B.1 summarizes the 22 trends and notable relationships between them.

For the sake of simplicity, Table B.1 focuses on the relationships most discussed during CJTFG meetings. Some very broad relationships (overarching many trends) are not shown in Table B.1—notably, the lack of business cases and use cases and established business processes for new technology and increasing pressures to add cybersecurity protections. These two wide-impact trends can be seen as negatively affecting virtually all technology-capability and technology-need trends.

Once we had a set of trends and relationships between them, we assembled them into a "social network" using the open-source social-network analysis tool NodeXL (NodeXL, undated). Doing so allowed us to visualize the full network of how the trends related to one another. It also allows for conducting various analyses on

- further grouping the 22 trends into a handful of overarching themes on how technologies are affecting criminal justice, going from *what kinds of trends there were* (as in Table B.1) to *what they were collectively trying to tell us*
- identifying which trends had the most relationships, either directly or indirectly, and thus are *most centrally involved in driving impacts on criminal justice.*

Table B.1
Trends and Relationships from the CJTFG

Trend	Notable Relationship to Other Trends
1. Technology can have both unintended consequences and unanticipated benefits.	• Having core use cases (2) could reduce risk but also hamper novel uses of technology. • Planners should consider unanticipated consequences of the emerging technology trends, including analytics and big data (12), situational-awareness displays (13), touch and rapid-DNA (14), and remote weapon-detection systems (15). • Researchers are seeing unanticipated consequences in terms of evidence-storage demands, legal demands, and public and law-enforcement behavioral changes related to BWCs (21) and supporting digital-evidence management (20).
2. Core business and use cases and other key reference material for emerging technologies are lacking.	
3. Established business processes for operationalizing new technologies are lacking.	• Core use cases (2) are a key part of establishing business processes.
4. Awareness of existing training and reference material on criminal-justice technologies is lacking.	
5. The reality of have and have-not agencies is a barrier to shared criminal-justice capabilities.	• Barrier to nationwide information-sharing (17) and enterprise-wide integration (18) • Barrier to court officers being able to handle large volumes of video (20)
6. Legal foundations and case law for new surveillance technologies are lacking.	• Further complicates increasing political and social pressures on surveillance technologies (9)
7. Pressure is increasing to move toward guardianship, with competing pressures to crack down on violence and terrorism.	
8. Pressure is increasing for law-enforcement technology to focus on accountability.	• Pressure to move toward guardianship (7) is a major driver of this trend. • Demand for accountability is a driver of the pressures related to surveillance-related technologies (9). • This is a major driver of demand for BWCs (21).
9. Pressure is increasing to address issues that consistently affect agencies' use of new surveillance technologies.	• Affects analytics conducted on surveillance-related data (13) • Affects situational-awareness capabilities when displaying surveillance-related data (14) • Affects potential use of new DNA-analysis systems (15) • Affects potential use of remote weapon-detection capabilities (16) • Affects how criminal justice–wide information integration applications might occur (18) • Affects the use of BWCs (21)
10. Pressure is increasing to have cybersecurity protections.	• General lack of awareness of existing material on improving cybersecurity (example of 4) • Demand for cyber protections is an increasing pressure on surveillance technologies (9)
11. Pressure is increasing to investigate cybercrimes.	• General lack of awareness of existing material on conducting cyber investigations (key example of 4) • Closely connected with increasing pressures for cybersecurity protections (10)
12. Going dark means that increasing commercial pressures on unbreakable encryption are hampering criminal investigations.	• The rise of strong encryption is driven, in part, by pressures for cybersecurity protections (10), as well as pressures against government surveillance (9) • Complicated by a lack of legal foundations and case law (6)

Table B.1—Continued

Trend	Notable Relationship to Other Trends
13. Analytics are emerging, and use of big data is being enabled.	• Lack of awareness of what can be done (example of 4) • Lack of nationwide law-enforcement information-sharing (17) • Criminal justice–wide information integration (18)
14. Situational-awareness displays, enabling devices, and data streams are emerging.	• Lack of awareness of what can be done (4) • Enabled by real-time analytics capabilities (13) • Lack of nationwide law-enforcement information-sharing (17)
15. Touch and rapid-DNA systems are emerging.	
16. Remote weapon-detection capabilities are emerging	
17. Law enforcement needs to advance from small-scale to nationwide information-sharing.	• Heavily dependent on having core business cases to show operational value of sharing (2)
18. Integration needs to enable new models of criminal justice across the enterprise.	• Heavily dependent on having core business cases to show the operational value of integration (2)
19. The value of fusion centers needs to be defined and disseminated.	• Need to have core business cases that show their operational value (2) • Could be a key resource for training on crime and counter-terrorism analysis (4)
20. Digital-evidence management needs to be supported on a massive scale.	
21. Fielding of BWCs needs to be supported on a large scale.	• Major driver of massive-scale digital-evidence management (20)
22. Less-lethal weapons are needed.	• In part driven by demands to move toward guardianship models of policing (7)

NOTE: The numbers in parentheses correspond to the trend numbers. So a *2* in parentheses indicates that the trend in the current row is related to trend 2 in the way described in the current bullet.

Figure B.3 graphs the relationships between the trends, showing their social network. It also assigns the trends into clusters, each of which represents an underlying theme.

Identifying Themes

We sorted the trends into clusters via the node-grouping feature in NodeXL, which adapts clustering algorithms from the Stanford Network Analysis Platform (Leskovec and Sosič, 2016). The trends within a cluster are comparably densely related to one another and less densely connected to trends outside the clusters. Each cluster implies an underlying *key theme* linking them together. The six clusters, and their underlying themes, were (clockwise from the top left in Figure B.3)

1. *emergence of big data and analytics—but also challenges using them.* This group connects the emergence of big data and analytics, situational-awareness displays, cyber investigations, and the information-sharing needed to use them effectively. The group also brings in the lack of awareness of technology resources and training, which directly hamper information-sharing and analysis.

Figure B.3
Relationships Among Emerging Trends

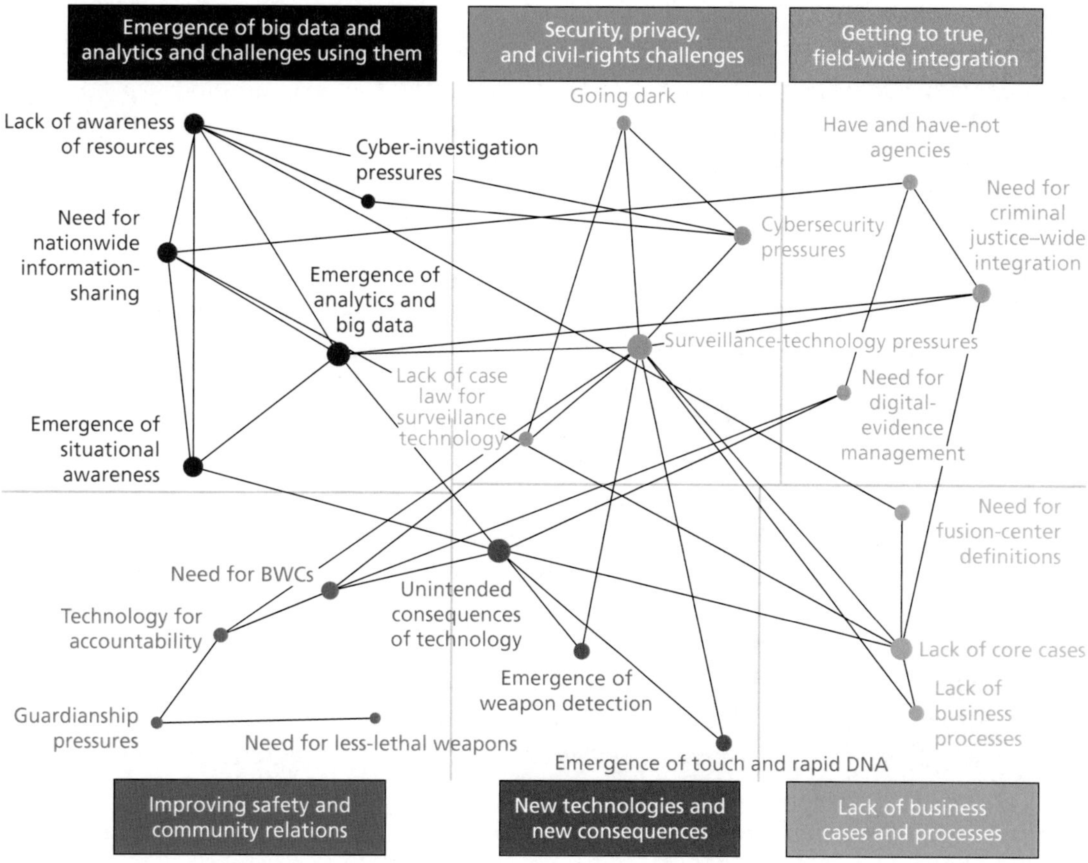

NOTE: Larger circles indicate more-central nodes.
RAND *RR1987-B.3*

2. *surveillance, privacy, and cybersecurity challenges.* This cluster brings together surveillance trends (most predominantly, social pressures, such as privacy concerns and lack of legal foundations related to surveillance technologies) and cybersecurity.
3. *getting to true, field-wide integration.* This cluster brings together the need for criminal justice–wide information integration with a major example of the need for integration (digital-evidence management), combined with a major barrier (have versus have-not agencies).
4. *improving safety and community relations.* This cluster connects trends related to societal pressures on guardianship and using technology to improve safety and accountability—most notably, the strong demands for BWCs and improved less-lethal options.
5. *new technologies and unanticipated consequences.* This group brings together the unanticipated consequences of new technologies with two emerging technologies (DNA and weapon detection).
6. *lack of core cases and processes.* This cluster brings together the lack of core business cases with the lack of core business processes, using the need for cases defining the value of fusion centers as an example.

Identifying Which Themes Are Most Central

The size of the node for each trend in Figure B.1 reflects a measure of *centrality*—that is, a measure of how interconnected the trends are to other trends.[1] Trends with a high degree of centrality are worth further exploration and consideration because improvements and changes related to them are likely to influence multiple trends and lead to additional impacts. That said, centrality does not equal the group's perception of importance; a trend can have few relationships with other trends yet still be held to be critical by the CJTFG. As shown in Figure B.3, the trends that were most central to the CJTFG's discussions (in the social-network analysis sense) were

- *pressures to address issues that consistently affect agencies' use of new surveillance technologies*—notably, security, privacy, and civil-rights policies. This trend linked to every surveillance-related trend (both sensors and analytics). Panelists also discussed that specific issues include demands for cybersecurity protections and demands to tightly define how the technologies would be used (i.e., core use cases).
- The *emergence of analytics and big data* had implications for making sense out of all the new data being collected and shared across agencies (i.e., providing agencies). As noted, it also had surveillance policy implications; panelists also saw it as being hampered by a lack of awareness among practitioners of what capabilities were available and how to get them.
- The *lack of core business cases/use cases* for emerging technologies hampered the use of some technologies, especially information sharing–related applications; it also complicated policy issues related to surveillance technologies.
- Panelists saw *unanticipated consequences*, both negative and positive, as important to consider for the emerging technology trends.
- A few additional themes were not as central as the first four, but worth noting:
  - Like the emergence of analytics and big data, the *emergence of situational-awareness displays and devices* similarly offered potential, major benefits in helping make sense of data about the world, but also potential unintended consequences, such as causing information overload.
  - Similarly, panelists saw the emerging needs for *criminal-justice enterprise–wide information integration* and *nationwide information-sharing* as core necessary conditions to make use of the new data to be collected and analyzed; they also saw them as hampered by a lack of use cases defining what information most needs to be shared, surveillance-related policy pressures (because much of said information is surveillance-related), and a lack of awareness of how to share information.
  - The *lack of awareness of existing training and reference material on criminal-justice technologies* broadly hampered adoption and use of a variety of technologies and made those technologies more vulnerable to attack.

Figure B.4 summarizes the relationships between trends that cut across the six thematic clusters. The single biggest set of interconnections is between the lack of business cases and

[1] Here, the measure used is *eigenvector centrality* (Newman, 2008), which can be roughly summarized as rewarding trends that both have many links to other trends and have links to other trends that themselves have many links.

Figure B.4
Interconnectedness Between Criminal Justice Technology Forecasting Group Themes

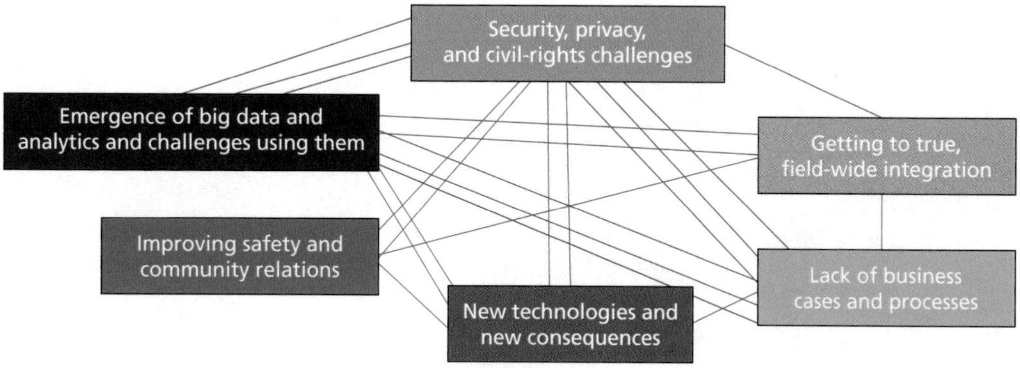

RAND *RR1987-B.4*

processes; the emergence of big data, analytics, and associated challenges; and security, privacy, and civil-rights challenges.

Analytic Summary

Figure B.5 provides a summary view of the trends and themes, in mind-map format. The four most central trends are noted in red flags; the next four most central trends are marked with yellow flags. The dashed red lines and their label indicate the major set of intertheme relationships. The core narrative with these relationships is that IT opportunities predominate in the current environment, but exploiting those opportunities is hampered by a lack of business case and business-process knowledge on how to use them, as well as a lack of knowledge on needed security, privacy, and civil-rights protections.

Identifying Recommendations and Action Items

In the weeks following CJTFG meeting 3, BJA, in conjunction with the RAND team, identified several near-term action items for the CJTFG to respond to some of the pressing trends and impacts. BJA tasked Global to cosponsor two task teams to improve information-sharing of health and other nontraditional criminal-justice data with service providers outside the criminal-justice system. BJA also tasked RAND to develop a concept of operations for a federally sponsored crime-analysis capability.

Identifying and Developing Recommendations

The RAND team reviewed the notes from the CJTFG's meetings, focusing on panelists' ideas for responses to the emerging trends. From analyzing the discussions of potential responses to trends, we identified a set of candidate recommendations that met the following criteria:

- *specific*, in terms of the discussion indicating a defined way ahead that implies

Figure B.5
Emerging Themes and Constituent Trends from the CJTFG

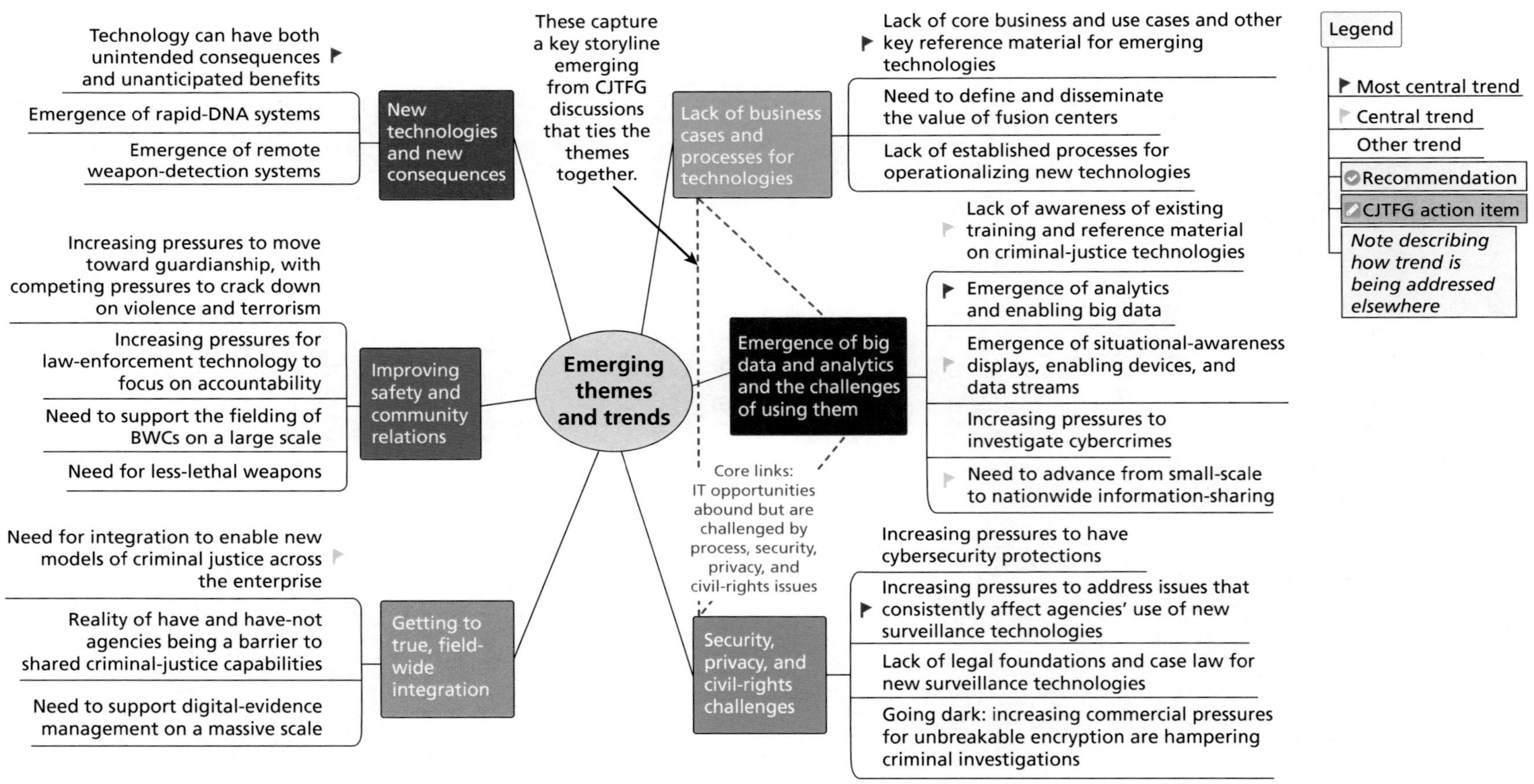

NOTE: Colored rectangles are themes. Outlined items are trends.

RAND *RR1987-B.5*

  - a specific set of actions to take
  - a party to take those actions, whether funders, researchers, practitioner and technical associations, or some combination
  - a logical relationship as to how carrying out the way ahead would help mitigate the problem (or take advantage of the opportunity)
- *substantive*, offering the possibility to gain significant benefits to criminal-justice practice in the next three to five years, and
  - are novel solutions, going beyond recommendations that had been typically heard in the past (BJA mentioned not wanting to hear "we need better leadership on information-sharing" again, as one example)
  - have a basis in research evidence, if available
- *comprehensive*, collectively covering all of the themes and most of the constituent trends
  - For the few trends not covered, the CJTFG and the RAND team identified ways in which outside initiatives are addressing these trends.
  - Each trend is addressed by at least one recommendation, action item, or initiative outside the CJTFG.
- *nonredundant*, in terms of not being addressed through other major initiatives
  - For example, we did not identify recommendations regarding cybersecurity, digital-evidence management, or general support for fielding BWCs because these are topics that others are addressing.

In contrast, CJTFG discussions that were mostly rhetorical, mostly asked about a point of information, or lacked a defined way ahead did not become candidate recommendations.

We identified a list of ten candidate recommendations as follows, sorted by theme:

- lack of established business processes for new technologies
  - Start work on developing and disseminating a common but tailorable business process for acquiring new technologies. Existing BJA and association resources related to technology strategy and acquisition can be leveraged here.
- lack of awareness of existing training and reference materials
  - Start research and experimentation on improving how technology resource materials are organized (within and across major reference sites), how technology material is presented, and how new resources are advertised.
- the reality of have and have-not agencies
  - Identify and disseminate examples of regional models for sharing capabilities, then develop promising practices based on these examples.
- growing and directly competing pressures to "demilitarize" and "crack down" on violence and terrorism
  - Identify those combinations of criminal-justice practices and supporting technologies that offer the greatest potential for effectiveness both in reducing crime and terrorism and in improving community relations.
- increasing pressures affecting the use of surveillance technologies
  - Work with communication experts to develop ways to educate the public on how criminal-justice technologies work (or do not work) in the real world.
- going dark

  - Collect quantitative data on the extent of the problem and investigative solutions to strong encryptions.
- potential emergence of remote weapon-detection systems
  - Assess the real-world practicality of these systems both now and over the next few years; if they are promising, study when these systems might be most appropriately used from legal and crime-prevention perspectives.
- need for information integration to enable new models of criminal justice
  - Challenge new methods to change cultures to better support information-sharing and safeguarding.
- need for digital-evidence management on a massive scale
  - Assess whether CJIS (or another federal agency) should establish a federally sponsored cloud capability for video and other digital evidence.
- need to enable fielding of BWCs
  - Facilitate exchanges between the United States and the UK on lessons learned about using cameras for both investigative and accountability purposes.

Meeting 4 (2016) was a structured review of the analyses described above and, primarily, the draft recommendations. Prior to the meeting, we circulated a memo listing the recommendations. During the meeting, we reviewed each recommendation and had the panelists provide detailed feedback on what they thought of the validity and importance of each and whether they had any additional context to provide. We also asked whether they had any additional recommendations to respond to the trends and overarching themes. The discussion led to consolidating and dropping some of the recommendations, refining others, and creating and endorsing two additional recommendations.

Meeting 5 (2017), finally, included a second review of the refined recommendations and actions, as well as a review of this report.

Concept for a Federally Sponsored Crime-Analysis Capability

Overview

BJA's CJTFG recommends the creation of a multisource, multijurisdiction crime-analysis environment that would be freely available to law-enforcement agencies that regularly upload data to NIBRS. The environment would be hosted in a cloud data center and would rely primarily on open-source and government-owned tools. The environment would (1) host a suite of analytic tools that are often too difficult or time-consuming for smaller agencies to set up, (2) simplify the process of sharing crime and other incident data across agencies, and (3) facilitate the conversion from UCR summary reporting to NIBRS.

Background

BJA has established the CJTFG to assess the major technology-related trends that are most likely to affect criminal-justice practice significantly in the next three to five years and identify potential responses to leverage emerging opportunities and mitigate emerging threats. In its January 2015 meeting, the CJTFG noted several linked trends:

- The FBI is transitioning all crime incident reporting from UCR summary statistics to NIBRS incident reporting by 2021. Although the major practitioner associations strongly support the transition, CJTFG members have heard of substantial concerns about the transition, including whether the cost of upgrading agencies' systems to report data about individual incidents outweighs a perceived limited benefit to individual agencies.
- The NIBRS data are sufficiently fine-grained and detailed—including fields on modi operandi of crimes, for example, as well as specific addresses and time and date stamps—to support a range of crime analyses. These include hot-spot policing, geospatial predictive policing, and crime linking. If the NIBRS data can be further combined with entity data (i.e., data about offenders, victims, vehicles and locations—e.g., the data uploaded to N-DEx, LInX, or COPLINK), the combined data could be used to identify lists of potential suspects for unsolved crimes (and clusters of crimes).
- A variety of IT trends are making it much easier to consolidate large volumes of data from multiple sources, analyze the data, and provide operationally useful displays to practitioners as a result. These include continued migration toward cloud and shared-service solutions for storing large volumes of data; migration toward open-source analytic and programming tools (e.g., R, Python, MySQL); and the emergence of mash-ups, which are web applications that display operationally useful information to users employing data

from multiple sources. The City of Chicago's open data portal, for example, offers a wide range of citywide data sets that can be displayed in tabular form or on a map, including all crimes since 2001 (City of Chicago, undated).

Collectively, these trends imply an opportunity to provide key crime-analysis capabilities to law-enforcement agencies using NIBRS data (and perhaps entity data) for limited effort, providing agencies with a local operational benefit for converting to NIBRS.

Concept

Figure C.1 shows the concept for the proposed multisource, multiagency crime-analysis environment.

Agencies would upload NIBRS-compliant incident data and N-DEx entity data to the environment[1] and then would be able to see map displays that highlight crime hot spots, high-volume locations for incidents, and potentially linked crime series—notably including current hot spots and times, projected hot spots and times over varying time horizons (next shift to next year) and scales ("hot" building up from "hot street corner" through "hot neighborhood"), and potential crime clusters (likely by the same offender or group of offenders) that cut across jurisdictional boundaries, enabled by the multiagency information-sharing. Agencies would also have an option to share their data in the form of crime maps and charts with the public.

If entity data are uploaded, agencies would be able to see social-network diagrams and tables that identify potential suspects for unsolved crimes. Network diagrams are an existing capability in N-DEx that could be leveraged using a mash-up approach. Suspect identification would be a new feature that would identify statistical-similarity matches between crime details

Figure C.1
Proposed Crime-Analysis Environment

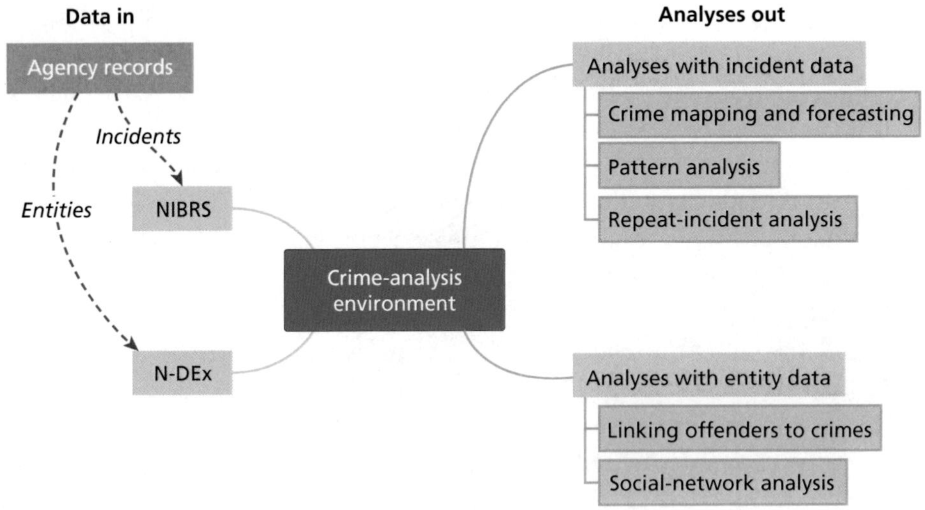

RAND *RR1987-C.1*

[1] The environment might also permit uploading data in LInX, COPLINK, or other formats yet to be determined.

(e.g., time, space, modi operandi, physical descriptions) and offenders' prior histories. It draws from earlier manual and machine-learning approaches that require the integration of incident details and criminal history information.

This description is preliminary: Specific capabilities are to be determined in consultations with law-enforcement personnel. From a technical perspective, the environment is tentatively slated to be a secure web portal that serves as a mash-up of NIBRS, N-DEx existing query and analytics capabilities, mapping (e.g., OpenStreetMap), and other data services to provide displays of interest to analysts.

Access control and other security features are also to be determined. We currently envision that the environment would be hosted on a CJIS policy–compliant cloud. We also envision that agencies would have a great deal of flexibility in setting access-control requirements for themselves, other agencies, and the public.

Next Steps

Presuming that there is an interested federal sponsor, the next step would be to work with law-enforcement practitioners and technical experts to create a detailed specification for the proposed environment. We propose working with both practitioner associations (e.g., chiefs' and sheriffs' associations, crime analysts' associations) and developer groups supported by the federal government (IJIS Institute, Global, Standards Coordinating Council) to work out the operational functions and technical details.

Bibliography

Aliens, James Cameron, director, 20th Century Fox, 1986.

"Amazon Echo," *Amazon.com*, undated. As of March 30, 2016:
http://www.amazon.com/Amazon-SK705DI-Echo/dp/B00X4WHP5E

Andrews, David A., Stuart William Harmer, Nicholas J. Bowring, Nacer D. Rezgui, and Matthew J. Southgate, "Active Millimeter Wave Sensor for Standoff Concealed Threat Detection," *IEEE Sensors Journal*, Vol. 13, No. 12, December 2013, pp. 4948–4954.

ARJIS—*See* Automated Regional Justice Information System.

Automated Regional Justice Information System, home page, undated. As of March 30, 2016:
http://www.arjis.org/SitePages/Home.aspx

BJA—*See* Bureau of Justice Assistance.

Bureau of Justice Assistance, "Body-Worn Camera Toolkit," undated. As of January 3, 2017:
https://www.bja.gov/bwc/

Casey, Pamela M., Jennifer K. Elek, Roger K. Warren, Fred Cheesman, Matt Kleiman, and Brian Ostrom, *Offender Risk and Needs Assessment Instruments: A Primer for Courts*, Williamsburg, Va.: National Center for State Courts, 2014. As of May 30, 2017:
http://www.ncsc.org/~/media/microsites/files/csi/bja%20rna%20final%20report_combined%20files%208-22-14.ashx

Center for Internet Security, "MS-ISAC: Multi-State Information Sharing and Analysis Center," undated. As of March 29, 2016:
https://msisac.cisecurity.org/

City of Chicago, "Chicago Data Portal," undated. As of August 8, 2017:
https://data.cityofchicago.org/

Data and Civil Rights, "Data and Civil Rights: A New Era of Policing and Justice," website for conference, Washington, D.C., October 27, 2015, c. 2015. As of January 3, 2017:
http://www.datacivilrights.org/2015/

DHS—*See* U.S. Department of Homeland Security.

FBI—*See* Federal Bureau of Investigation.

Federal Bureau of Investigation, "National Data Exchange (N-DEx) System," undated (a). As of March 30, 2016:
https://www.fbi.gov/about-us/cjis/n-dex

Federal Bureau of Investigation, "UCR Program Data Collections," undated (b). As of March 30, 2016:
https://www.fbi.gov/about-us/cjis/ucr/ucr-program-data-collections

Garvie, Clare, Alvaro Bedoya, and Jonathan Frankle, *The Perpetual Line-Up: Unregulated Police Face Recognition in America*, Washington, D.C.: Georgetown Law Center on Privacy and Technology, October 18, 2016. As of January 5, 2017:
https://www.perpetuallineup.org/

Glenn, Jerry, "Futurizing Teaching vs. Futures Courses," *Social Science Record*, Vol. 9, No. 3, Spring 1972, pp. 26–29.

Global Standards Council's Justice-to-Health Services Task Team, "Aligning Justice-to-Health Priority Exchanges Task Team Final Report," Global Justice Information Sharing Initiative, August 2014. As of August 17, 2017:
https://it.ojp.gov/GIST/171/Aligning-Justice-to-Health-Priority-Exchanges-Task-Team-Final-Report

Global Strategic Solutions Working Group, "Prioritizing Justice-to-Health Exchanges Task Team Final Report, Bureau of Justice Assistance, U.S. Department of Justice, March 2014. As of August 17, 2017:
https://it.ojp.gov/GIST/167/Prioritizing-Justice-to-Health-Exchanges-Task-Team-Final-Report

Goodman-Delahunty, Jane, and Hielkje Verbrugge, "Reality, Fantasy and the Truth About CSI Effects," *InPsych*, August 2010. As of May 30, 2017:
http://www.psychology.org.au/publications/inpsych/2010/august/goodman/

Hollywood, John S., Dulani Woods, Richard Silberglitt, and Brian A. Jackson, *Using Future Internet Technologies to Strengthen Criminal Justice*, Santa Monica, Calif.: RAND Corporation, RR-928-NIJ, 2015. As of August 17, 2017:
https://www.rand.org/pubs/research_reports/RR928.html

IACP—*See* International Association of Chiefs of Police.

IBM, "i2 COPLINK," undated.

IJIS Institute, "Springboard Certification," undated. As of March 30, 2016:
http://www.ijis.org/?page=Springboard

International Association of Chiefs of Police, *IACP Technology Policy Framework*, January 2014. As of March 29, 2016:
http://www.theiacp.org/Portals/0/documents/pdfs/
IACP%20Technology%20Policy%20Framework%20January%202014%20Final.pdf

Jackson, Brian A., and Dulani Woods, *Criminal Justice Technology Taxonomy Web Tool*, Santa Monica, Calif.: RAND Corporation, TL-158-NIJ, 2015. As of March 13, 2016:
http://www.rand.org/pubs/tools/TL158.html

Justice Information Sharing, Bureau of Justice Assistance, Office of Justice Programs, U.S. Department of Justice, "Global Federated Identity and Privilege Management," undated. As of March 29, 2016:
http://www.it.ojp.gov/initiatives/gfipm

Kanowitz, Stephanie, "Police Video Redaction: Where Big Data Meets Privacy," *GCN*, March 18, 2016. As of March 30, 2016:
https://gcn.com/articles/2016/03/18/video-redaction-tech.aspx

Law Enforcement Cyber Center, home page, undated (a). As of March 29, 2016:
http://www.iacpcybercenter.org/

Law Enforcement Cyber Center, "Incident Reporting," undated (b). As of March 29, 2016:
http://www.iacpcybercenter.org/resource-center/incident-reporting/

Leskovec, Jure, and Rok Sosič, "SNAP: A General Purpose Network Analysis and Graph Mining Library," *ACM Transactions on Intelligent Systems and Technology (TIST)*, Vol. 8, No. 1, October 2016, Art. 1.

Matthies, Carl, and Tina Chiu, *Putting a Value on Crime Analysts: Consideration for Law Enforcement Executives*, New York: Vera Institute of Justice, March 27, 2014. As of August 4, 2017:
https://www.vera.org/publications/
putting-a-value-on-crime-analysts-considerations-for-law-enforcement-executives

McEwen, Harlin R., "ICAM Summit Final Report Announcement," *Information Sharing Environment*, blog, January 27, 2015. As of March 29, 2016:
https://www.ise.gov/blog/chief-harlin-r-mcewen/icam-summit-final-report-announcement

Minor, Joe, "Touch DNA: From the Crime Scene to the Crime Laboratory," *Forensic Magazine*, April 12, 2013. As of March 30, 2016:
http://www.forensicmag.com/articles/2013/04/touch-dna-crime-scene-crime-laboratory

National Information Exchange Model, "Cyber," undated.

National Network for Safe Communities at John Jay College, "Swift, Certain, and Fair," undated. As of March 30, 2016:
http://nnscommunities.org/our-work/strategy/swift-certain-fair

Newman, M. E. J., "Mathematics of Networks," in Steven N. Durlauf and Lawrence E. Blume, eds., *The New Palgrave Dictionary of Economics*, 2nd ed., Palgrave Macmillan, 2008, pp. 1–12.

NIEM—*See* National Information Exchange Model.

NodeXL, Social Media Research Foundation, undated. As of August 4, 2017:
http://www.smrfoundation.org/nodexl/

Northrop Grumman, "Law Enforcement Information Exchange (LInX)," undated. As of March 29, 2016:
http://www.northropgrumman.com/Capabilities/PublicSafety/Pages/LawEnforcementInformationExchange.aspx

Open Source Policing, "Open Source Policing," undated. As of January 3, 2017:
http://opensourcepolicing.org/

OpenStreetMap, "OpenStreetMap Powers Map Data on Thousands of Web Sites, Mobile Apps, and Hardware Devices," undated. As of March 30, 2016:
https://www.openstreetmap.org/about

"Our Approach to Privacy," *Apple.com*, undated. As of March 29, 2016:
http://www.apple.com/privacy/approach-to-privacy/

Periscope, home page, undated. As of March 29, 2016:
https://www.periscope.tv/

Perry, Walter L., Brian McInnis, Carter C. Price, Susan Smith, and John S. Hollywood, *Predictive Policing: Forecasting Crime for Law Enforcement*, Santa Monica, Calif.: RAND Corporation, RB-9735-NIJ, 2013. As of November 13, 2014:
http://www.rand.org/pubs/research_briefs/RB9735.html

Pinola, Melanie, "How to Use Mind Maps to Unleash Your Brain's Creativity and Potential," *Lifehacker*, September 19, 2013. As of January 3, 2017:
http://lifehacker.com/how-to-use-mind-maps-to-unleash-your-brains-creativity-1348869811

Pollack, Andrew, "Building a Face, and a Case, on DNA," *New York Times*, February 23, 2015. As of March 30, 2016:
http://www.nytimes.com/2015/02/24/science/building-face-and-a-case-on-dna.html

President's Task Force on 21st Century Policing, *Final Report of the President's Task Force on 21st Century Policing*, Washington, D.C.: Office of Community Oriented Policing Services, 2015. As of August 4, 2017:
https://cops.usdoj.gov/pdf/taskforce/taskforce_finalreport.pdf

Public Law 104-191, Health Insurance Portability and Accountability Act of 1996, August 21, 1996. As of August 7, 2017:
https://www.gpo.gov/fdsys/pkg/PLAW-104publ191

Public Law 114-113, Consolidated Appropriations Act of 2016, December 18, 2015. As of August 8, 2017:
https://www.gpo.gov/fdsys/pkg/PLAW-114publ113/content-detail.html

Python Software Foundation, "Python Is Powerful . . . and Fast," undated. As of March 30, 2016:
https://www.python.org/about/

Rahr, Sue, and Stephen K. Rice, "From Warriors to Guardians: Recommitting American Police Culture to Democratic Ideals," *New Perspectives in Policing*, Washington, D.C.: U.S. Department of Justice, National Institute of Justice, NCJ 248654, April 2015. As of August 4, 2017:
https://www.ncjrs.gov/pdffiles1/nij/248654.pdf

R Foundation, "The R Project for Statistical Computing," undated. As of March 30, 2016:
https://www.r-project.org/

Sirkin, Harold L., Perry Keenan, and Alan Jackson, "The Hard Side of Change Management," *Harvard Business Review*, October 2005. As of March 29, 2015:
https://hbr.org/2005/10/the-hard-side-of-change-management

Tolson, Mike, "Chief Justice Roberts: Technology Among Top Issues for Court," *Houston Chronicle*, October 17, 2012. As of March 29, 2016:
http://www.chron.com/news/houston-texas/houston/article/High-court-not-so-politicized-Chief-Justice-3957626.php

U.S. Computer Emergency Readiness Team, "Automated Indicator Sharing (AIS)," undated. As of March 29, 2016:
https://www.us-cert.gov/ais

U.S. Department of Homeland Security, *Privacy Impact Assessment for the Acquisition and Use of License Plate Reader Data from a Commercial Service*, DHS/ICE/PIA-039, March 19, 2015. As of January 5, 2017:
https://www.dhs.gov/sites/default/files/publications/privacy-pia-ice-lpr-march2015.pdf

U.S. Senate, Committee on Homeland Security and Government Affairs, Permanent Subcommittee on Investigations, *Federal Support for and Involvement in State and Local Fusion Centers: Majority and Minority Staff Report*, Washington, D.C., October 3, 2012. As of August 4, 2017:
https://www.hsgac.senate.gov/download/report_federal-support-for-and-involvement-in-state-and-local-fusions-centers

Xmind 7.5 (v3.6.51), June 29, 2016. As of August 17, 2017:
http://www.xmind.net/download/previous/

Zapotosky, Matt, "East Coast Rapist Indicted in Pr. George's, the Place Where He Said His Crimes Began," *Washington Post*, March 26, 2013. As of March 29, 2016:
https://www.washingtonpost.com/local/east-coast-rapist-indicted-in-pr-georges-the-place-where-his-crimes-began/2013/03/26/d3b346ce-963e-11e2-894a-b984cbdff2e6_story.html

Zapyon, "File:Futures wheel 01.svg," January 13, 2009. As of January 3, 2017:
https://en.wikipedia.org/w/index.php?curid=8613285